Columbia University

Contributions to Education

Teachers College Series

No. 483

AMS PRESS
NEW YORK

PROBLEM SOLVING IN ARITHMETIC

A STUDY OF CERTAIN FACTORS IN THE ALLOCATION OF ARITHMETIC PROBLEMS INVOLVING ONE, TWO, THREE, AND FOUR STEPS OR PROCESSES IN REASONING

By

LEON NELSON NEULEN, Ph.D.

TEACHERS COLLEGE, COLUMBIA UNIVERSITY
CONTRIBUTIONS TO EDUCATION, No. 483

BUREAU OF PUBLICATIONS
Teachers College, Columbia University
NEW YORK CITY
1931

Library of Congress Cataloging in Publication Data

Neulen, Leon Nelson, 1894-
Problem solving in arithmetic.

Reprint of the 1931 ed., issued in series: Teachers College, Columbia University. Contributions to education, no. 483.
Originally presented as the author's thesis, Columbia.
Bibliography: p.
1. Arithmetic--Study and teaching. 2. Arithmetic--Problems, exercises etc. I. Title. II. Series: Columbia University. Teachers College. Contributions to education, no. 483.
QA135.5.N36 1972 372.7'2 70-177116
ISBN 0-404-55483-0

Reprinted by Special Arrangement with Teachers College Press, New York, New York

From the edition of 1931, New York
First AMS edition published in 1972
Manufactured in the United States

AMS PRESS, INC.
NEW YORK, N. Y. 10003

ACKNOWLEDGMENTS

The author desires to take this opportunity to acknowledge his appreciation of the encouragement received from Dean William F. Russell and Professors Thomas Alexander, Paul R. Mort, R. G. Reynolds, and George D. Strayer. To Professors John R. Clark, J. R. McGaughy, and C. B. Upton he is grateful for valuable aid and counsel.

L. N. N.

CONTENTS

TABLES, CHARTS, AND GRAPHS

TABLES

CHARTS

GRAPHS

PROBLEM SOLVING IN ARITHMETIC

CHAPTER I

INTRODUCTION *

Previous Studies

During the past five years there have been a number of investigations relating to problem solving in arithmetic. The first studies that were made concerned themselves largely with the elimination of obsolete problems and the attempt to make verbal problems more practical. Five such studies are cited in the monograph written by Hydle and Clapp [16]. Since that time several studies have been made which concerned themselves with the question of how pupils solve reasoning problems; [1] specific causes of failure and success in problem solving; [2] and techniques for improving the solution of arithmetic problems.[3] Thus far, no investigations have been made to determine the relationship, if any, between the number of "steps" † in a problem and its difficulty. Nor has any study been made to discover what constitutes current practice in the grade placement of "steps" in problem solving.

Certain studies have been made pointing out some of the factors which influence problem solving, and a few studies have established definite relationships between ability in arithmetic fundamentals and ability in problem solving. But not all these studies agree; for example, Lutes [23], Osburn [36], Stevenson [42], and Wilson [50] agree that one of the chief factors of failure in problem solving is due to inability to compute accurately. Morton found a high correlation between computa-

* Note: Numbers appearing in parentheses refer to the numerical order in which the reference may be found in the Bibliography.

[1] For studies supporting this topic, see Bibliography, references 7, 8, 21, 24, 37.

[2] For studies supporting this topic, see Bibliography, references 1, 16, 18, 25, 30, 34, 35, 38, 39, 41, 47.

[3] For studies supporting this topic see Bibliography, references 6, 12, 22, 23, 31, 33 (p. 589), 42, 46, 48.

† Note: The number of "steps" refers to the number of operations necessary to arrive at the solution.

tional ability and problem-solving ability;[4] and the first tests in Winch's [51] experiment showed a positive correlation between ability in reasoning and accuracy in numerical computation (coefficients, .69 to .85). Curtis and Stone,[5] on the other hand, found (1) that there is no correlation between accuracy in the mechanical phases of arithmetic and reasoning; and (2) that there is but little correlation between accuracy in the combinations as such and in problems involving the same combinations. Stone [43] concluded in his study that "reasoning ability in fundamentals implies ability in English." Myers [24] found a closer relationship between reasoning ability and general intelligence than between reasoning ability and skill in computation. This same conclusion was reached by Morton.

To date, no one has attempted to determine the ability of beginning third, fourth, fifth, sixth, and seventh grade pupils to solve problems graded with respect to the number of steps involved in their solution.

Purpose of the Present Study

The purpose of the present investigation is threefold:

1. To analyze courses of study in order to discover what constitutes current practice in the grade placement of problems with respect to the number of steps involved.

2. To determine the ability of beginning pupils in the third, fourth, fifth, sixth, and seventh grades to get correct answers to written problems involving one, two, three, and four steps.

3. To determine relationships between several factors involved in problem solving; these include reading ability, computational skill, and intelligence.

Subject Defined

In this experiment the writer has not attempted to measure both reasoning capacity and ability in fundamentals by means of the same problem tests. The study measures, primarily, "the ability to earn scores on tests of ability to solve verbal problems, which is the school's definition of arithmetical reasoning."[6]

Many school administrators judge the ability of pupils to solve problems wholly in terms of their ability to get the right answers.

[4] *The Elementary School Journal,* June 1926, p. 750.
[5] Hydle and Clapp, reference 16, pp. 7 and 8.
[6] National Council of Teachers of Mathematics, *Second Yearbook,* p. 51.

To such administrators a problem is correctly solved if the right answer is obtained. It is incorrect if a wrong answer is obtained, even though the reasoning is correct. Such a point of view is regarded by many teachers as extreme, since the getting of the wrong answer to a problem is in no sense an indication that the pupil cannot reason correctly in problem solving.

In this study the ability to solve problems is measured also in terms of right or wrong answers. The procedure followed in this study differs, however, from that mentioned above in that an attempt has been made to eliminate to a reasonable extent the computational element; this was accomplished as follows: The pupils were first given the Preliminary Test. This was immediately followed by the Inventory Test which contained all the types of computations found in the Preliminary Test. The pupils were then drilled upon their computational weaknesses as disclosed in the Inventory Test; the assumption was that these computational weaknesses would thus be reasonably remedied. The Final Test, which was exactly the same as the Preliminary Test, was then given. Since the Final Test contained only those types of computation upon which drill had recently been given, it was *assumed* that inability to compute correctly was eliminated to a reasonable extent in the Final Test. Hence in the Final Test a right answer was taken as an indication of ability to solve a problem correctly, and a wrong answer as evidence of inability to solve it. This entire study, therefore, is based upon the assumption [7] that after drill has been given on computational difficulties, one may measure the ability to solve problems solely in terms of right or wrong answers. This assumption is a more reasonable one than that usually made by school men, who assume, without making any attempt whatever to eliminate the computational element, that a wrong answer indicates lack of ability to solve a problem.

It should also be clearly understood that this investigation includes a study of certain factors *in* and not *for* the allocation of arithmetic problems involving one, two, three, and four steps or processes in reasoning.

No attempt has been made to determine the grade placement of problems with respect to the number of "steps" involved. The study concerns itself primarily with problem work as it is

[7] See Appendix for further discussion of this assumption.

related to current practice. It is not the chief purpose of this study to determine the actual reasoning ability of elementary school children in Grades 3 to 7 inclusive, nor to discover the factors which should determine the special allocation of the number of steps in problem solving. Neither is it the province of the present study to construct a series of ideal reasoning problem tests, other than those which may be found in currently used instructional materials for Grades 3, 4, and 5. The province of this study, however, comprises the following aims:

1. To analyze courses of study in order to discover what constitutes current practice in the grade placement of problems with respect to the number of steps involved.

2. To determine the ability of beginning pupils in third, fourth, fifth, sixth, and seventh grades to get correct answers to written problems involving one, two, three, and four steps.

3. To discover what correlation exists between reading ability and arithmetic errors in problem solving which is not explained entirely by computational difficulties.

4. To discover an index of mastery in problem solving based upon errors due entirely to erratic performance.

5. To determine the correlation that exists between intelligence levels and problem-solving ability.

6. To compare the relationship between problem-solving ability and general intelligence with that between problem-solving ability and skill in computation.

7. To determine whether "dependable habits of computation" seem essential to success in problem solving.

8. To discover whether improvement in computational accuracy increases ability to earn scores on verbal problem tests.

9. To determine the correlation that exists between intelligence test scores and gains made between the preliminary tests and the final tests in problem solving.

Sources of Data

A review of the literature on problem solving brought to light varying opinions relating to the grade placement of reasoning problems in arithmetic and certain relationships existing in problem-solving ability. An investigation was therefore undertaken in order to clarify the issues. The approach has been largely statistical and experimental. The principal sources of data have included courses of study, test materials similar to those found in currently used instructional materials, and test results from two widely separated school systems that comprise approximately three thousand pupils.

Techniques Used

The purpose throughout this study has been essentially to assemble and present all the statistical facts bearing specifically on the problem as stated. Whenever interpretations have been made, they have been little more than a recapitulation of the data themselves. In large measure, therefore, the study is objective. The techniques used may be summarized as follows:

1. Testing. Preliminary, inventory, and final tests.
2. Drilling. A high degree of mastery of computational skills.
3. Analysis from analytical record sheets.
4. Statistical treatment and interpretation. A three-level grouping for each grade based on M.A. and I.Q.

CHAPTER II

PROCEDURE

General Statement

Twenty-four outstanding courses of study were analyzed to discover the agreement or disagreement concerning the grade placement of the number of "steps" in problems in Grades 1 to 6 inclusive. Paragraph comprehension tests, selected from The New Stanford Achievement Test, and the Otis Group Intelligence Tests, were given by the superintendent of each school system to all pupils at the close of the school year 1928-29. Preliminary inventory and final tests were administered as follows: The pupils were first given a preliminary test. This was immediately followed by an inventory test which contained all the types of computations found in the preliminary test. The pupils were then drilled upon their computational weaknesses as disclosed in the inventory test until a mastery (eight out of ten responses correct) was acquired. The final test, which was exactly the same as the preliminary test, was then given. The preliminary and the inventory tests were given at the close of the third week of school in the fall of 1929. The final tests were given at the close of the necessary computational drill period, which varied, with different teachers, from six to eight weeks. In giving the tests there were none of the distractions usually incidental to the intrusion of a stranger in a classroom. Each teacher gave her own pupils the problem tests. Both oral instructions and detailed explanations were given for administering and scoring the tests. The teachers were cautioned to avoid certain practices which might influence the results of the tests and were asked to follow their written instructions very closely. In scoring the papers all the problems that were not attempted or that were incomplete were counted wrong. In recording the results only those pupils were considered who had completed all units in all the tests. The pupils were from two widely separated school systems. The experiment was explained to both teachers and principals. All knew that the preliminary tests were to be repeated at the close of six or eight weeks and all agreed not to explain the tests or in any way coach their pupils in problem-solving work during the period of experimentation.

The Tests Used

Three types of tests were used: preliminary, inventory, and final.

The *preliminary tests* comprised a series of test problems for each of Grades 3, 4, and 5, similar in nature and difficulty to those found in currently used instructional materials for these grades. Each series consisted of twenty-five problems for each of the one-, two-, three-, and four-step difficulties in problem solving, with the exception of the series used in the third grade. This series consisted of twenty-five problems for each test of one- and two-step difficulty only. The series of tests constructed for the fifth-grade pupils were given also to the sixth- and seventh-grade pupils. This afforded a basis for observing and comparing growth in problem solving from year to year in the upper grade levels.

The *inventory tests,* which were given immediately following the preliminary tests, contained the exact regularly arranged computations involved in the preliminary tests, except that the examples were presented in reverse order. These inventory tests determined the computational weaknesses to be drilled upon.

The *final tests* were the same as the preliminary tests.

Drill Required

Each teacher was to drill all students upon the computational weaknesses revealed in the inventory tests until a high degree of mastery (eight out of ten responses correct) should be acquired. A short computational drill period of from ten to fifteen minutes was set aside each day from the regular arithmetic period in order to give the pupils incidental drill on errors that needed correction. The amount of time required to complete the drill work varied, with different teachers, from six to eight weeks. The final tests were given at the close of the required drill period.

Pupils Tested

Pupils in the beginning third, fourth, fifth, sixth, and seventh grades were selected for the experiment. More than two thousand elementary pupils of the Champaign, Illinois, and Teaneck, New Jersey, schools were tested. In order to maintain a consistent, complete performance, only pupils who had completed or attempted all the problems in all the tests were recorded on the final record sheets. The distribution by grades of pupils having complete records is shown above. These schools represent a reasonable sampling of children from urban life: two of the schools in Champaign, Illinois, are composed entirely of Negro children, and two schools in each city are composed of children from the poorer sections of the city; the larger group of pupils are from homes of professional men and tradespeople; and in one or two schools in each city the pupils came from wealthy homes. Hence the schools represented in these two cities would seem to furnish a fair sampling of urban school children.

Number of Pupils, per Grade, Having Complete Records, upon Which the Investigation Is Based

Grade 3	274
Grade 4	270
Grade 5	270
Grade 6	394
Grade 7	316
	1524

Instructions Given for Administering and Scoring the Tests

In order to secure more intelligent coöperation from the teachers, a special meeting was called at which all teachers were present. The purpose of the experiment was explained to them; oral instructions and detailed explanations were given for administering and scoring the tests, and the teachers were cautioned to avoid certain practices which might influence the results of the tests, and were asked to follow their instructions very closely.

Instructions

The mimeographed instructions sent out with each set of tests were as follows:

PRELIMINARY TESTS

Directions to Teachers of Grades 3, 4, 5, 6, and 7

1. A series of Arithmetic Problem Tests have been devised for the purpose of determining the ability of grade pupils to solve one-, two-, three-, and four-step reasoning problems. These tests have been prepared in mimeographed units which in themselves form a complete test.
2. Several units of the Inventory Tests should be given every day, preferably at the beginning of your regular arithmetic period and possibly also at the beginning of one or two other morning periods.
3. The teacher's judgment should in the last analysis be the sole criterion for judging the time limit of each test. When you consider that adequate time has been given for the solution of each unit of problems, collect the papers, mark them, record the results on the class record sheet, and bind them together with the other units which go to make up the complete set of twenty-five problems.
4. No results should be recorded on the class record sheet unless *all units* for each complete test of twenty-five problems have been worked or attempted.
5. Encourage every pupil in your class to attempt or complete *all units* for the entire series.
6. Record the results according to the directions set forth on your class record sheet.
7. After all results have been recorded, stack the papers together, group all the papers for each complete set of twenty-five problems together, and return the material to the central office.

Answer Sheet

An answer sheet was mimeographed and sent out with each full set of problems. A sample sheet appears below.

ANSWER KEY FOR GRADE 4

One-Step		*Two-Step*		*Three-Step*		*Four-Step*	
Prob. No.	Ans.	Prob. No.	Ans.	Prob. No.	Ans.	Prob. No.	Ans.
1	2	1	20¢	1	1¢	1	15
2	$1.06	2	18	2	20	2	27
3	86¢	3	9	3	36	3	24¢
4	7	4	Yes	4	14	4	10¢
5	$1.32	5	68	5	40	5	$1.30
6	Yes	6	55¢	6	28	6	45
7	34	7	17	7	$1.66	7	$4.71
8	33	8	118	8	6¢	8	$6.65
9	$1.15	9	11	9	50¢	9	$1.54
10	$12.70	10	$5.76	10	$4.86	10	$345.95
11	65¢	11	$1.02	11	55¢	11	$7.85
12	6	12	$1.00	12	3	12	$2.15
13	5	13	5¢	13	294	13	$78.00

14	87	14	24¢	14	19¢	14	$6.02
15	500	15	$16.00	15	$7.70	15	42¢
16	624	16	250	16	25¢	16	20
17	496	17	15	17	13	17	5
18	24	18	10	18	$2.02	18	3
19	12	19	36	19	18	19	$2.12
20	17	20	$1.60	20	40¢	20	9½
21	25¢	21	Yes	21	$704.00	21	28
22	$50.00	22	6	22	31	22	$9.35
23	91¢	23	$16.75	23	30¢	23	2.20
24	8¢	24	8 lb. 14 oz.	24	51	24	5
25	45¢	25	2	25	72	25	52¢

(Record all problems *that are wrong* on your class record sheet in accordance with directions.)

INVENTORY TESTS

Directions to Teachers of Grades 3, 4, 5, 6, and 7

1. A series of Inventory Tests have been prepared to determine the computational difficulties involved in the one-, two-, three-, and four-step reasoning problems which were given to the pupils in the Preliminary Tests.
2. Several units of the Inventory Tests are to be given every day, preferably at the beginning of your regular arithmetic period and possibly also at the beginning of one or two other morning periods.
3. The teacher's judgment should determine the time limit of each test. When you consider that adequate time has been given for the solution of all the problems in the test, collect the papers, check any errors, and record the results on the class record sheet. If any computational error has been made on any *part* of a problem (some will have four parts if the problem is a four-step problem) record the entire problem as in error.
4. The problems of the Inventory Test are the reverse of those given in the Preliminary Test. This will also be seen by examining your class record sheet. In recording the results of your Inventory Tests, disregard the numbers and *use only the letters* on your class record sheet.
5. From your Inventory Tests, determine the exact type of computational errors to be drilled upon. Drill upon these weaknesses, *along with your other general drill work,* until you know through frequent checking that a high degree of mastery (eight out of ten responses correct) of these weaknesses has been acquired by each pupil. Then give the Final Tests. The drill will probably cover four or five weeks.[1]

Answer Sheet

The type of answer sheet which was mimeographed and sent out with each set of problems appears on opposite page.

[1] It should be noted that in some instances the actual drill period extended from six to eight weeks instead of four or five weeks.

Answer Sheet for Grade 4

Inventory Test, No. I (One Step)

(a)	45	(b)	8	(c)	91	(d)	$50.00	(e)	$.25
(f)	17	(g)	12	(h)	24	(i)	496	(j)	624
(k)	500								
(l)	87	(m)	5	(n)	6	(o)	65	(p)	$12.70
(q)	$1.15	(r)	33	(s)	34	(t)	.85	(u)	$1.32
(v)	7	(w)	86	(x)	1.06	(y)	2		

See that the answers in the first row correspond to the right problems in the Inventory Test. Check the *wrong problems.* Fold down the upper row of answers and use the second row the same way, etc.

Remember to record the results on your class record sheet *according to letters* and not according to number sequence.

The directions for the final tests read as follows:

FINAL TESTS

Directions to Teachers of Grades 3, 4, 5, 6, and 7

1. The Final Tests are the same as those used in the Preliminary Tests.
2. Give the Final Tests the same way as the Preliminary Tests, that is, at the beginning of your regular arithmetic period and also at the beginning of one or two other morning periods.
3. The teacher's judgment should in the last analysis be the criteria for judging the time limit of each test. When you consider that adequate time has been given for the solution of each unit or problems, collect the papers, mark them and record the *wrong problems* on the class record sheet as directed.
4. No results should be recorded on the class record sheet unless *all units* for each complete test of twenty-five problems have been worked or attempted.
5. Every pupil who completed the other test units should try to complete all the units in the Final Test.
6. After all results have been recorded, stack the papers together, grouping together all the papers for each complete set of twenty-five problems and return the material, *together with your class record sheets,* to the central office.

The answer sheet was the same as that used for the Preliminary Tests.

Class Record Sheet

A class record sheet (see Chart I) was devised for recording and analyzing the problems that were either missed or solved incorrectly. Directions appeared at the top of the sheet.

CHART I

Class Record Sheet of Problem-Solving Results

C = Mark all problems *correct* in the Preliminary Test with a "C." (black-blue ink.)

/ = Mark all problems *that are wrong* in the Inventory Test with a diagonal *red mark* from right to left, corner to corner.

\ = Mark all problems *that are wrong,* in the Final Test, with a diagonal *red mark* from left to right, corner to corner.

Er. C. D. = Errors due entirely to computational difficulties. /

Er. Ap. I = Errors that are apparently inherent, not due to computational difficulties. \

Er. CD&I = Errors that may be due to both computational and inherent difficulties. ×

Erratic = Errors due to erratic performance. ☐ ⊘ ⧅ ☒

P. T. = Preliminary Test.

F. T. = Final Test.

School.................... Class....................

Grade.................... Teacher....................

ONE-STEP PROBLEMS

Problems of Inventory Test			y	x	w	v	u	t	s	r	q	p	o	n	m	l	k	j	i	h	g	f	e	d	c	b	a	Total Correct			Final Check-up			
M.A.	I.Q.	Names	1	2	3	4	5	6	7	8	9	10	11	12	13	14	15	16	17	18	19	20	21	22	23	24	25	P.T.	F.T.	Gains	Er. C.D.	Er. Ap. I	Er. CD&I	Erratic

In a general teachers' meeting the use of the Class Record Sheet was demonstrated and explained in full. The columns labeled P.C. and I.Q.* were filled in at the central office with the raw scores made in each, before they were sent out to the teachers. In this way it was possible to have each class ranked according to I.Q. before the teachers received and recorded their test results—a procedure which also facilitated the final tabulation.

In order to provide an accurate check on the recording of all class scores, each teacher was asked to check each pupil's work by the following formula: "The sum of all numbers recorded in the last four columns called 'Final Check-Up,' plus the actual number of 'C's' appearing in the columns labeled 'One-Step Problems' with no cross marks on them, should total 25." All Class Record Sheets were rechecked in the central office in order to make sure that this formula was carried out and that the results checked accurately.

Final Tabulation

For purposes of final tabulation, the class results for each grade were recorded in sequence according to I.Q. on the following record form. (See Chart II on following page.)

The Final Record Form is to be read as follows: In column I, pupil A made two errors in all of his one-step problems (sub-column 1) which may be charged to errors due to other causes than purely computational causes.† He made no errors in any of his two-step problems which could be charged to either computational causes or other causes. In his three-step problems (sub-column 3) he made four errors that were entirely due to causes other than computation. In his four-step problems he made eleven errors that were entirely due to errors other than computational errors. This made a total of seventeen errors in all his one-, two-, three-, and four-step problems which were entirely due to causes other than computation. In column II, pupil A made a score of ninety-eight on the paragraph comprehension test in the New Stanford Achievement tests. In column III, pupil A's mental age received a score of 153, as determined by the Otis Group Intelligence tests. In column IV, pupil A solved

* P.C. stands for paragraph comprehension; I.Q. means intelligence quotient.

† This was determined by subtracting the number of problems correct on the preliminary test (see column IV, 1) from the total number of problems which he worked or attempted (25). Then from this result again subtracting the errors which were due entirely to computational difficulties (see column VIII, 1, "C").

CHART II

Final Record Form

Pupil	I					II	III		IV				V				VI			
	Errors not Explained by Computational Dif.					Par. Comp. Score	Intel. Score		Score on Preliminary Test				Score on Final Test				Gains			
	1	2	3	4	Tot.		M.A.	I.Q.	1	2	3	4	1	2	3	4	1	2	3	4
A	2	0	4	11	17	98	153	133	22	25	18	11	25	23	17	16	3	−2	−1	5
B	2	4	7	9	22	92	149	133	23	20	16	16	25	25	22	21	2	5	6	5
C	3	0	6	8	17	100	154	131	22	24	17	16	25	24	21	19	3	0	4	3
D	1	6	3	10	20	96	153	128	24	18	20	15	23	19	18	9	−1	1	−2	−6
E	0	4	3	3	10	100	150	126	25	20	21	19	25	25	22	20	0	5	1	1

twenty-two out of the twenty-five one-step (sub-column 1) problems; all the twenty-five two-step problems; eighteen three-step problems; and eleven four-step problems. After six to eight weeks of remedial instruction on the computational weaknesses revealed in the inventory tests, we find in column IV that in the final problem tests, this same pupil A worked all the one-step problems correctly; twenty-three of the twenty-five two-step problems correctly; and seventeen of the four-step problems correctly. In column VI we find a record of the differences expressed between columns IV and V. In the final tests, pupil A gained or made an increase of three problems worked correctly in the one-step problem tests; he lacked two problems of doing as well on the two-step problems; he lacked one problem of doing as well on the three-step problem tests; but he succeeded in working five more of the four-step problems correctly in the final tests than he did in the preliminary tests. From a correction of the inventory tests, we find in column VII that pupil A solved correctly, or, in other words, had no computational errors in twenty-two of the one-step problems; eighteen of the two-step problems; eighteen of the three-step problems; and twenty of the four-step problems. In column VIII, we find that pupil A had only one error that was due to computational difficulty, no errors that were due to inherent difficulties, no errors that were due to both computational and inherent difficulties, and four errors that were due to erratic performances in the one-step problems (sub-column 1); no errors due to computational, inherent, computational and inherent difficulties, and eight errors due to erratic performance in the two-step problems (sub-column 2); three errors due to computational difficulty, no errors due to inherent difficulties, one error due to computational and inherent difficulties, and eleven errors due to erratic performance in the three-step problems (sub-column 3); three errors due to computational difficulties, six errors due

CHART II—(*Continued*)

FINAL RECORD FORM

VII				VIII															
No. Correct in Inventory Test				Problem Errors due to															
				1				2				3				4			
1	2	3	4	C.	I.	C&I.	E.	C.	I.	C&I.	E.	C.	I.	C&I.	E.	C.	I.	C&I.	E.
22	18	18	20	1	0	0	4	0	0	0	8	3	0	1	11	3	6	0	9
23	23	22	24	0	0	0	4	1	0	0	9	2	0	0	11	0	1	0	12
24	18	22	22	0	0	0	3	1	0	0	8	2	1	0	10	0	1	0	6
23	19	18	9	0	0	0	6	1	1	0	13	2	0	1	12	0	8	1	8
25	22	23	18	0	0	0	0	1	0	0	6	1	1	0	5	3	1	0	6

to inherent difficulties, no errors due to both computational and inherent difficulties, and nine errors due to erratic performance in the four-step problems (sub-column 4). Similar details are to be read for each pupil, showing the various factors in problem solving which were tabulated for purposes of study.

As each pupil's results were recorded on the Final Record Form, a red line was drawn through all his scores on the Class Record Sheet. This facilitated recording and locating of the scores desired, and avoided a duplication of results previously recorded.

When this classification of errors had been completed, each grade was arbitrarily divided according to I.Q. scores into three groups. In order that the two extremes would not result in small numbers of pupils and thus enhance the Probable Error, a division was made as follows: those having I.Q. scores of 116 and upward were considered the bright or upper group; those having I.Q. scores ranging from 100 to 115 inclusive were considered the average or middle group; and those having I.Q. scores of 99 or less were considered the backward or lower group.

To insure accuracy in the transfer of the results from the Class Record Sheets to the Final Record Forms, grand totals were determined for each group and checked, wherever possible.[2]

[2] In column I, the sub-columns 1, 2, 3 and 4 should equal the "Total" column; in column VI, the sub-column 1 should equal column V, sub. 1, less column IV, sub. 1, etc.

CHAPTER III

AN ANALYSIS OF COURSES OF STUDY IN ARITHMETIC

The author was aware of the fact that a wide range of practice existed in allocating to the different grade levels the number of steps to be taught in reasoning problems; for example, in the school system of Ithaca, N. Y., it is advocated that all problems involving two steps of reasoning should be deferred until the pupils reach the fifth grade; Baltimore County, on the other hand, advocates the introduction of two-step problems in the second grade.

TABLE 1

A SUMMARY OF RESULTS IN A SURVEY OF RECENT COURSES OF STUDY AS TO STEP-PROBLEMS RECOMMENDED FOR DIFFERENT ELEMENTARY GRADES

	GRADE					
	1	2	3	4	5	6
Number of courses included in the study	20	23	23	24	24	24
Number of courses in which *no problems are prescribed* for the given grade	11	0	0	0	0	0
Number of courses in which 1-step problems only are prescribed or are used in the course of study	8	20	19	3	0	0
Number of courses in which 1- and 2-step problems only are prescribed or used in the course of study	0	1	4	21	13	0
Number of courses in which 3-step problems are prescribed in the course	0	0	0	0	10	10
Number of courses in which no limit is placed	0	0	0	0	1	14
Number of courses in which limitations are placed upon the number of steps in problems for different ability groups	0	0	1	2	1	2

This table should be read: There were 20 courses of study examined for Grade 1; 23 courses for Grades 2 and 3; and 24 courses for Grades 5 and 6. Out of the 20 courses examined for Grade 1, 11 prescribed no problem work, and 8 prescribed one-step problems only. Out of 23 courses of study examined for Grade 2, 20 prescribed one-step problems only, and 1 prescribed both one and two-step problems only, etc.

Accordingly, in order to discover the degree of agreement or disagreement of current grade practice in the use of the number of "steps" in problem solving, twenty-four of the outstanding, recent courses of study in arithmetic, as determined by the Bureau of Curriculum Research, Teachers College, Columbia University, were analyzed. Two of these courses of study were state courses—those of Louisiana and Montana; and two were county courses—those of Baltimore County and Garrett County, Md. The remaining twenty courses of study were selected from sixteen different states, in order that they might be as representative as possible of general arithmetic practice in the United States. It should be noted also that only twenty courses were found for first-grade work; and twenty-three courses for second- and third-grade work.

Grade Placement of Step-Problems

A detailed analysis of this study is set forth in Table 2. The summary of results appears in Table 1.

One-Step Problems

From Table 1, it will be observed that whereas eight of the twenty courses of study state that only one-step problems shall be used in the first grade, three of the twenty-four courses of study advocate that one-step problems should be limited to the fourth grade only. Between these limits we find that twenty of the twenty-three courses prescribe one-step problems for second grade only, while nineteen of the twenty-three courses prescribe one-step problems for third grade only. This situation seemingly substantiates the following statement: "In general, teachers believe that one-step problems should predominate in the second and third grades, reserving two- and three-step reasoning for the later grades." [1]

Two-Step Problems

The range of two-step problem work in courses of study extends from Grade 2 to Grade 5 inclusive, with the major emphasis in Grades 4 and 5. It should be noted that twenty-one of the twenty-four courses of study limit the fourth grades to one- and two-step problems, while thirteen of the twenty-four

[1] Department of Superintendence, *Third Yearbook, National Education Association,* p. 95. Washington, D. C., 1925.

TABLE 2

SURVEY OF RECENT COURSES OF STUDY AS TO STEP-PROBLEMS RECOMMENDED FOR DIFFERENT ELEMENTARY GRADES

	Grade					
	1	2	3	4	5	6
1. Kansas City, Mo. 1928	No problems	1-step	1-step with 2-step toward end of year	1- and 2-step	1- and 2-step not stated but problems studied	No statement but evidently not limited
2. Denver, Colo. 1926	No problems	1-step	1-step	1- and 2-step	1-, 2-, and 3-step	1-, 2-, and 3-step
3. Seattle, Wash. 1928 for 3d, 5th, 6th 1927 for 4th	No problems		1-step	1- and 2-step	No statement but evidently not limited	No statement but evidently not limited
4. San Diego, Cal. 1928	No problems	No statement in course but 1-step problems used	1-step	1- and 2-step	1- and 2-step	Oral 1-step written not limited
5. Rochester, N. Y. 1923	No course of study	No statement in course but 1-step problems given	1-step	1- and 2-step	1- and 2-step	No statement but evidently not limited
6. Grand Rapids, Mich. 1928	No course of study	1-step	1-step	1- and 2-step	1-, 2-, and 3- step	1-, 2-, and 3-step
7. Houston, Tex. 1927–28	Oral problems without numbers	1-step oral	1-step	1- and 2-step	1-, 2-, and 3-step	6th grade in the Junior High School. Does not set limits in course
8. Cleveland, Ohio 1927–28	Simple problems suggested	Simple problems suggested	1-step for average groups. 1- and 2-step for superior groups	1- and 2-step	1 and 2-step	No statement but evidently not limited

9. Ithaca, N. Y. 1927	No problems	1-step	1-step	1-step	1- and 2-step	1-, 2-, and 3-step
10. Mitchell, S. D. 1928	No problems	1-step	1-step	1- and 2-step	1-, 2-, and 3-step	1-, 2-, and 3-step
11. St. Louis, Mo. 1926	No problems	1-step	1-step	1- and 2-step	1- and 2-step	No statement
12. Winston-Salem, N. C. 1928	Select process for 1-step problems	Select process for 1-step problems	1-step	1- and 2-step	1-, 2-, and 3-step as shown by problems used	1-, 2-, and 3-step as shown by problems used
13. Portland, Me.	No problems	No statement in course but simple 1-step problems used	1-step	1- and 2-step	1- and 2-step	No statement but evidently not limited
14. Baltimore, Md.	No problems	1-step oral	1-step	1- and 2-step	1- and 2-step	No statement but evidently not limited
15. Worcester, Mass.	No problems	No statement in course but 1-step problems used	No statement in course but 1-step problems used	1-step	1- and 2-step	No statement but evidently not limited
16. Summit, N. J.	No problems	1-step	1-step	1- and 2-step	1- and 2-step	No statement but evidently not limited
17. Passaic, N. J.	No course of study	No statement but 1-step problems used	No statement but 1-step problems used	1-step for slow groups, 1- and 2-step for average and superior groups	1- and 2-step for slow and average group. 2- and 3-step for superior groups	2-step for slow groups. 3-step for average and superior groups
18. Wilmington, Del	Not stated	"Simple oral problems"	1-step oral	1- and 2-step	1- and 2-step	No statement but evidently not limited
19. Trenton, N. J. 1928	1-step	1-step	1-step	1-step for slow groups, 1- and 2-step for others	1- and 2-step	1- and 2-step for slow and average groups. 1,- 2-, and 3-step for superior groups

TABLE 2—(*Continued*)

SURVEY OF RECENT COURSES OF STUDY AS TO STEP-PROBLEMS RECOMMENDED FOR DIFFERENT ELEMENTARY GRADES

	GRADE					
	1	2	3	4	5	6
20. Cicero, Ill. 1928	No course of study	No course of study	No course of study	2-step	2- and 3-step	2- and 3-step
21. Baltimore Co., Md.	Not stated but 1-step problems used	1-and 2-step	1-and 2-step	1- and 2-step	1-, 2-, and 3-step	1-,2-, and 3-step
22. Garrett Co., Md. 1926–28	1-step oral	1-step oral	1-step	1- and 2-step	1-,2-, and 3-step	"Several steps"
23. Montana State	1-step	1-step	1-step	1-step	1- and 2-step	No statement. Evidently not limited
24. Louisiana State	1-step	1-step	1-and 2-step	1- and 2-step	1-, 2-, and 3-step	1-, 2-, and 3-step

courses of study limit the fifth grades to one- and two-step problems only. This does not appear to be in agreement with the statement of practice made by Dr. Brueckner when he says: "Usually the work in problem solving in the second, third, and most of the fourth grade is limited to one-step problems." [2] A similar practice is implied in the following statement: "Of 488 problems in a widely used textbook in arithmetic for Grades 3 and 4, 404 were 'one-step' problems and 73 were 'two-step' problems." [3] If other "widely used" textbooks organize their one-step problem-solving work around the third and fourth grades, there would appear to be a wide gap between "theory and practice," that is, between what is prescribed in courses of study and what is actually practiced in textbook work.

Three-Step Problems

Regarding three-step problems, there appears to prevail an equally divided opinion. Ten of the twenty-four courses prescribe three-step problems for the fifth grade only, while ten prescribe three-step problems for the sixth grade only. Problems of more than two steps appear to be introduced or prescribed in either the fifth or the sixth grade. Fourteen of the twenty-four courses of study for the sixth grade place no limit on the number of steps which may be taught at that time. It would appear, therefore, that in these courses of study three-step problems, at least, are formally introduced before the sixth grade. This does not support Dr. Brueckner's statement: "Problems of more than two steps are formally introduced in grades six to eight." [4]

Problems of More Than Three Steps

While only one course of study, that of Seattle, Washington, places no limitation on the number of steps to be taught in the fifth grade, fourteen of the twenty-four courses of study remove the limitation requirements in the sixth grade. Teachers from the sixth grade upward may apparently teach as many problem steps as they choose.

The range of opinion regarding the required placement of step-

[2] Brueckner, L. J., "The Supervision of Arithmetic." *The Supervision of Elementary Subjects*, p. 35, D. Appleton and Company, 1929.

[3] *Third Yearbook, Department of Superintendence*, p. 95, 1925.

[4] Brueckner, L. J., *op. cit.*

problems is considerable not only from grade to grade, but also within a grade. Of the twenty-four courses of study examined for Grade 5, thirteen limited the grade to one- and two-step problems; ten to three-step problems; and one had no limitations whatever.

Summary

A summary analysis of twenty-four outstanding courses of study in arithmetic reveals a wide variation of practice in the grade placement of one or more reasoning steps in problem solving—not only between grades but also within a single grade.

The allocation of one-step problems ranges from Grades 1 to 4; the allocation of two-step problems ranges from Grades 2 to 5.

The greatest variation for a single grade in prescribed problem steps appears in Grade 5. One course of study has no limit to the number of steps that may be taught in the fifth grade. Ten courses of study would limit the fifth grade to three-step problems only; and thirteen courses of study would limit this grade to only one- and two-step problems.

One-step problems are largely allocated to the second or the third grade; two-step problems to the fourth grade; three-step problems to the fifth or the sixth grade; and more than three-step problems to the sixth grade.

There appears to be no close agreement among the outstanding courses of study with respect to the grade placement of problems that involve one or more reasoning steps.

CHAPTER IV

CONSTRUCTION OF TESTS USED

Preliminary and Final Tests

A series of problem tests were constructed for each of Grades 3, 4, and 5, similar in nature and difficulty to those found in currently used instructional materials for these respective grades. Each series consisted of twenty-five problems each of one-, two-, three-, and four-step difficulty, with the exception of the series used in the third grade. This third-grade series consisted only of twenty-five problems of one-step difficulty and twenty-five problems of two-step difficulty.

The twenty-five problems in each series were broken up into smaller groups or units of problems in order that they might be distributed over several short periods of work during the morning, and thus prevent undue fatigue on the part of the pupils who were solving them. This plan also made easier the work of the teachers in administering and scoring the tests.

The manner in which the problem tests for each of the grades were divided into units or small problem groups is presented in the following tabulation:

Grade 3	One-step	7 units of 3 problems each 1 unit of 4 problems
	Two-step	11 units of 2 problems each 1 unit of 3 problems
Grade 4	One-step	5 units of 5 problems each
	Two-step	5 units of 4 problems each 1 unit of 5 problems
	Three-step	7 units of 3 problems each 1 unit of 4 problems

Grade 5 (cont.)	Four-step	11 units of 2 problems each 1 unit of 3 problems
Grade 5	One-step	2 units of 8 problems each 1 unit of 9 problems
	Two-step	3 units of 6 problems each 1 unit of 7 problems
	Three-step	5 units of 4 problems each 1 unit of 5 problems
	Four-step	7 units of 3 problems each 1 unit of 4 problems

Following are examples of the first two units or problems taken from the one-step series written for Grade 3.

ARITHMETIC REASONING PROBLEMS

GRADE 3

One-Step

School...................... Class......... Name...................

There is a space below each problem for figuring. Write your answer after each problem or on the line, whenever you find one. When you have finished working all of the problems, lay your pencil down and fold your arms.

1. Patty is 3 years old. Peter said, "I am 7 years older than Patty." How old is Peter?
2. Don counted 4 blue jays, 3 robins, and 5 sparrows, from his window. He said, "I see .... birds out here."
3. Leon helped Uncle Luke plant 225 cabbage plants. Of the plants, 213 lived, but .... died.

ARITHMETIC REASONING PROBLEMS

GRADE 3

One-Step

School................. Class........... Name......................

There is a space below each problem for figuring. Write your answer after each problem or on the line, whenever you find one. When you have finished working all of the problems, lay your pencil down and fold your arms.

4. Mother asked me to wash and count her jelly glasses. I said, "There are 67 glasses, but 12 are cracked, so you can use only .... of them."
5. Doris has two book shelves of her own. On one she has 12 books and on the other 14. How many books has she in all?

6. Joe and Tony tried to pick 32 quarts of strawberries in an hour, but when the time was up they had only 28 quarts. How many more quarts did they need to pick before they would have picked 32 quarts?

The directions for each unit of problems state that due time allowance should be given for full deliberation and the solution of all problems. No pupil was placed at a disadvantage in working the problems, because the papers were not collected until they had finished their solutions. Each problem series, therefore, was a test of "achievement or power" rather than a test of speed in "reasoning or mental alertness."

Between each problem in every unit, sufficient space was provided for necessary computation; this device eliminated the disadvantage of working the problems on scrap paper. The guiding principle in the set-up of all units was to minimize, as far as possible, the labor and time necessary for the computations involved in reaching a solution.

Inventory Computational Tests

The inventory tests contained the exact orderly arranged computations involved in the preliminary and the final tests, with the exception that the problems were presented in reverse order and lettered according to the alphabet. The series of inventory tests for each grade were divided into the following:

Grade 3	One-step	1 unit of 10 problems
		1 unit of 15 problems
	Two-step	2 units of 9 problems
		1 unit of 7 problems

Grade 4	One-step	1 unit of 11 problems
		1 unit of 14 problems
	Two-step	1 unit of 7 problems
		1 unit of 8 problems
		1 unit of 10 problems
	Three-step	1 unit of 8 problems
		1 unit of 7 problems
		1 unit of 10 problems
	Four-step	2 units of 6 problems each
		1 unit of 5 problems
		1 unit of 8 problems

Grade 5	One-step	1 unit of 11 problems
		1 unit of 14 problems
	Two-step	2 units of 8 problems each
		1 unit of 9 problems
	Three-step	1 unit of 9 problems
		2 units of 8 problems each
	Four-step	1 unit of 5 problems
		2 units of 7 problems each
		1 unit of 6 problems

Following are examples of one-step and four-step units taken from the fifth-grade series:

ARITHMETIC REASONING PROBLEMS

GRADES 5, 6, AND 7

One-Step

School.................. Class............ Name.......................

There is a space below each problem for figuring. Write your answer after each problem or on the line, whenever you find one. When you have finished all the problems, lay your pencil down and fold your arms.

1. To-day Mother bought me a pair of socks for 25¢, a dress for $2.36, a coat for $8.94, and a pair of shoes for $3.50. What did all my new clothes cost?
2. Alan had a rope 15 feet long. He cut off 6 feet to tie on his sled. How long was the piece that was left?
3. On Saturday Father took us all to the circus. He bought 5 tickets and paid $1.05 for each one. How much did he pay for all the tickets?
4. Last summer there was no rain for 42 days. Mr. Farmer said, "We've had no rain for .... weeks."
5. Father bought a suit for $43.50 and an overcoat for $37.25. How much did he pay for both of them?
6. Uncle Fred sent me a baby alligator from Florida. It was 9 inches long when it came and now it is 17 inches long. How many inches has it grown since I received it?
7. Aunt Fanny bought skates for each of her 3 nephews for Christmas. She paid $2.60 a pair for the skates. How much did the skates cost her in all?
8. During vacation Walter worked for $8 a week and earned $72. How many weeks did he work?

ARITHMETIC REASONING PROBLEMS

GRADES 5, 6, AND 7

Four-Step

School.................. Class............ Name.......................

There is a space below each problem for figuring. Write your answer after each problem. When you have finished working all the problems, lay your pencil down and fold your arms.

1. Tom's hens laid 18 eggs each for 9 days. He sold these eggs for 49¢ a dozen. Half the money he got for eggs he spent for chicken feed. How much did Tom spend for chicken feed?
2. Uncle Roger had $5000 in the bank when he bought his new car. The price of it was $1250. The dealer said "Because you can pay cash, I will make the price of the car $50 less." Uncle Roger paid cash for the car. He turned in his old car and was allowed $325 for it. After he had bought the new car, he had to spend $56 to insure it. How much money did Uncle Roger have left in the bank after he had insured the new car?
3. Jim bought 8 gallons of gasoline at 18¢ a gallon and 2½ quarts of oil at 20¢ a quart. How much change did he receive from a 5-dollar bill?

ARITHMETIC REASONING PROBLEMS

GRADES 5, 6, AND 7

Four-Step

School.................. Class............ Name ....

There is a space below each problem for figuring. Write your answer after each problem or on the line, whenever you find one. When you have finished all the problems, lay your pencil down and fold your arms.

4. Mr. Farmer had 40 crates of strawberries yesterday. There are 32 quarts in a crate and he paid 2¢ a quart to have them picked. He paid 10¢ a crate to have them hauled to market. How many dollars did Mr. Farmer have to pay to have the berries picked and hauled?
5. Uncle Frank sold 4 ducks at $1.60 each and 4 turkeys at $4.90 each. With his money he bought seed corn at $2 a bushel. How many bushels of seed corn did Uncle Frank buy?
6. Our 4th-grade class gave our teacher a basket of fruit at Thanksgiving time. We paid 45¢ for the basket. We put in ½ dozen oranges at 48¢ a dozen, a package of dates at 15¢, 6 red apples at 2 for 5¢, and a 10-cent grapefruit. We tied a green ribbon, that cost 10¢, on the handle. The present cost us ......¢.

METHOD OF SELECTING AND CONSTRUCTING PROBLEMS

A collection was made of recent textbooks, drill books, and standardized test materials on reasoning problems. These were studied for the purpose of constructing test problems similar in nature and difficulty to those found in the currently used instructional material. After a preliminary draft of such problems had been constructed, the writer refined these problems with the following criteria [1] in mind.

[1] Criteria Nos. 1 and 3 are found in [32], p. 105. Criteria Nos. 2 and 4 are found in [16], pp. 26 and 54. Criteria No. 5 is found in [44], pp. 220-24. Criteria No. 6 is found in [45], pp. 341-360. Criteria No. 7 is found in [28], p. 270.

1. Simple vocabulary
2. Setting easily visualized
3. Familiar terms
4. No "non-essential" elements
5. Situations familiar to children
6. Rôle of practical usefulness
7. Interesting content

The refined problems were then submitted to arithmetic specialists [2] who checked the problems and suggested several revisions. The final draft was then mimeographed and distributed to the teachers. The tests are reproduced below, but are not available for general use.

ARITHMETIC REASONING PROBLEMS

Grade 3

One-Step

1. Patty is 3 years old. Peter said, "I am 7 years older than Patty." How old is Peter?
2. Don counted 4 blue jays, 3 robins, and 5 sparrows, from his window. He said, "I see .... birds out here."
3. Leon helped Uncle Luke plant 225 cabbage plants. Of the plants, 213 lived, but .... died.
4. Mother asked me to wash and count her jelly glasses. I said, "There are 67 glasses, but 12 are cracked, so you can use only ...... of them."
5. Doris has two book shelves of her own. On one she has 12 books and on the other 14. How many books has she?
6. Joe and Tony tried to pick 32 quarts of strawberries in an hour, but when the time was up they had only 28 quarts. How many more quarts did they need to pick before they would have picked 32 quarts?
7. Allan had 17¢ and Uncle Will gave him 8¢ more. How much did Allan have in all?
8. Alice weighed 47 pounds. She went to the seashore and gained 5 pounds. So she weighed ...... pounds when she came back.
9. It is 234 miles to Aunt Lena's house. We drove 115 miles and then stopped for lunch. I said, "We have ...... miles to go yet, Father."
10. Mother bought a toothbrush for 45¢. She gave the man a half-dollar and let me have the change. How much did I get?
11. Ben said, "I have more pencils than anyone else. I have 4 red ones, 2 blue ones, 3 yellow ones, 2 brown ones, and 5 green ones." How many pencils has he?
12. Vera made 24 glasses of jelly. Thirteen were apple jelly and the other ...... glasses were currant jelly.
13. Bess and Mother went on a long trip to Grandma's. They drove 111

[2] Clifford B. Upton and John R. Clark.

miles the first day, 124 miles the next day, and 132 miles the last day. How far away did Grandma live?

14. To-day Nick practiced on his violin 12 minutes before lunch, 15 minutes after lunch, and 12 minutes after school. How many minutes did he practice in all?
15. Uncle George has 46 rabbits; 17 of them are large and ...... are small.
16. Grandpa put 58 chickens and 19 turkeys together in a pen. How many were there in all?
17. Sue bounced a ball 127 times without missing. Sally bounced hers 119 times. She said, "I wonder how many more bounces I needed to make me even with Sue." Can you tell her?
18. Mark saw a wagon marked $2.69. He counted his money and found he had $2.36. He said, "I'll have to save up ......¢ before I can get that wagon."
19. My father has made three trips in an airplane. Once he flew 132 miles, another time 246 miles, and the last time 157 miles. How many miles did he fly in all?
20. Nancy dried the glasses and silver for Mother. There were 12 glasses, 25 spoons, 24 forks, and 13 knives. How many things in all did she dry?
21. Edward had 28 pumpkins in his patch. He sold 13 and fed the rest to his father's cow. How many did he feed to the cow?
22. Joan said, "There are 96 pages in my book. 32 pages have no pictures on them. How many pages must have pictures on them?"
23. Father said, "Try to find me a dozen screws." Ray found 8 and May found 4. Was that a dozen? Write "Yes" or "No" on this line ....
24. Arthur shot 22 arrows into a target and Earl shot 28. Earl said, "I'm .... ahead of you, Arthur."
25. Sam ran an errand for the grocer. In one of his pockets Sam had a quarter and in another pocket a dime. How many cents did he have in all?

Grade 3

Two-Step

1. Rose has 10¢. If she buys a 3-cent apple and a 4-cent orange how much change should she get?
2. Ned earned 18¢. He spent 3¢ for a pencil. Then he gave his sister 5¢. How much had he left?
3. Biddy is sitting on 15 eggs. She hatched out 6 chicks this morning and 3 this afternoon. How many eggs are left to hatch?
4. Nancy had 3¢ and earned 3¢ more. Father said, "I'll give you enough to make a dime." How much did he give her?
5. I gave Dan a dime and asked him to get me a 5-cent stamp and a 3-cent stamp. When he brought the stamps, he gave me ......¢ in change.
6. There are 30 children in the 2nd grade. Yesterday it rained and there were only 12 boys and 11 girls in my grade, so ...... children stayed home.

7. Father sent for 25 pine trees and 25 spruce trees. He planted 36 trees and let me have the rest. How many trees did I get?
8. Bobby had 20¢. He bought a 4-cent ball, a 6-cent boat, and a 5-cent top. He said, "I have ......¢ left to put in my bank."
9. Uncle William had 18 cows. He sold 2 black ones and 3 red ones. How many cows did Uncle William have left?
10. Rena practiced on the piano 12 minutes after lunch and 15 minutes before supper. How many more minutes would she have had to practice before she would have practiced half an hour?
11. Grace wants to buy a plant for Mother. It costs 65 cents. She saved 24¢ last week and 29¢ this week. How much more does she need?
12. Paul had 21 bantam chickens and sold 12 of them to Ralph. He said, "Ralph has ...... chickens more than I have, now."
13. Frank picked 6 quarts of strawberries and Fanny picked 5 quarts. Mother said she needed 18 quarts to can, so I picked the other ...... quarts for her.
14. Mr. Farmer raised 32 bushels of potatoes in one field and 55 bushels in another field. He sold 49 bushels. How many bushels did he have left to sell?
15. Father bought a can with 15 quarts of oil in it. He put 4 quarts in the car last week and 4 quarts in the tractor this week. There are ...... quarts left in the can.
16. Alan had a dime in one pocket and 8¢ in another. He lost a nickel. How much did he have left?
17. Aunt Kate gave me a dime and a nickel. I want to buy a game that costs 18¢. How much more do I need?
18. Bob had 20 tomatoes on 1 vine. He gave 6 to Aunt Mary, he sold a dozen to Mr. Black, and the family ate the other tomatoes. How many tomatoes did the family eat?
19. Alice sorted her sewing basket and found she had 5 spools of cotton thread and 4 spools of silk. Two spools were just alike. She gave one of those to Mother. How many spools of thread did she have left?
20. Last fall we planted 6 rose bushes. Two of them died. This spring we planted 4 more and all are alive, so we now have ...... rose bushes.
21. Ellen is 16 years old. I am half as old as Ellen. Baby is half as old as I am, so Baby must be ...... years old.
22. Edward has 12 words to learn to spell. He learned 6 before supper, 2 since supper, and he is studying the other ...... now.
23. Fred had 4 marbles. He bought a dozen more. Then he lost some through a hole in his pocket. He has only 13 left. How many did he lose?
24. Uncle George gave me a dime and a nickel. I bought an 8-cent box of cookies and had ......¢ left.
25. Henry planted 3 rows of carrots and 4 rows of beets but his goat ate up 2 whole rows so there are only ...... rows left.

ARITHMETIC REASONING PROBLEMS

Grade 4

One-Step

1. Uncle Tim wants to make a shelf. He wants it 7 inches wide, but all the boards he can find are 9 inches wide. They are ...... inches too wide.
2. Jerry is buying himself some tools. He has a hammer that cost 38¢, a saw that cost 55¢, and a file that cost 13¢. How much has he spent so far?
3. Ellen bought 2 books and paid 43 cents for each of them. How much did she pay for both of them?
4. Mother hung 14 of my socks on the line this morning. How many pairs is that?
5. Jane got $3.75 for her birthday. She bought a sweater for $2.43. She put the rest of her money into her bank. How much did she put into her bank?
6. David earned 60¢ mowing lawns. He earned 25¢ weeding the garden. He said, "I've earned 85¢." Had he?
7. Our class visited the City Bakery to-day. We went in two busses. Each bus took 17 children. How many children were there in our class?
8. Uncle Charles has 66 sheep. He said, "I have too many. I'm going to sell half my sheep and keep only ......"
9. Harry is earning money to buy skates that cost $2.75. He has earned $1.60 raking leaves. How much does he still need?
10. Frank's new suit cost $7.98, his shoes cost $3.45, and his cap cost $1.27. How much did his things cost in all?
11. Polly is making her garden. She has bought 13 packets of seed at 5¢ each. So she has spent ...... cents for seed.
12. Mother has 24 cake candles. How many cup cakes can she trim if she puts 4 candles on each cup cake?
13. Aunt Bell had 23 quarts of cherries. She canned 18 quarts and made the rest into pies. How many quarts did she use for pies?
14. Grandpa has 235 bushels of apples. He sold 148 bushels from the orchard and put the rest in his cellar. How many bushels did he put in the cellar?
15. To-day Father bought 285 baby chicks and in two weeks he will buy 215 more. If none die, he will then have ...... chicks.
16. Uncle Jim has to make long trips. Last week he rode 375 miles on the train and flew 249 miles in an airplane. How many miles did he ride in all?
17. If you were riding in an airplane that was going 124 miles an hour, how far away would you be in 4 hours?
18. My brother works 6 days a week. In 4 weeks he works ...... days.
19. The baker has sweet buns that he sells for 3¢ each. If I give him 36¢ I can get ...... buns.

20. Janet is reading 6 pages every day in her new story book. There are 102 pages. She said, "It will take me ...... days to read this book."
21. Glen bought a raincoat for $4.75. He handed the man $5.00 and said to himself, "I ought to get a quarter in change." Was he right?
22. Father puts money in the bank every Saturday night. This month he put in $14.25, $10.25, $13.50, and $12.00. How much did he put in the bank all together?
23. We use 13 loaves of bread a week. We pay 7¢ a loaf, so our bread costs us ......¢ a week.
24. Leon saw in the toy shop a set of 3 kites for 24¢. He said, "That means each kite costs ......¢."
25. The 9 boys in our baseball team walked to the park to play and rode home on the trolley. The fare is 5¢. How much did it cost for all of us?

Grade 4

Two-Step

1. Helen needed 50¢. She earned 15¢ washing dishes and 15¢ dusting, and Uncle Harry gave her the rest. How much did Uncle Harry give her?
2. Bob has 4 white rabbits and 5 brown ones. Peter has twice as many rabbits as Bob. So Peter must have ...... rabbits.
3. Mrs. White picked 8 red roses and 10 pink ones. If she gave half her roses to Mother, how many roses did Mother get?
4. We had 24 hens and sold 7 that were not laying eggs. Then we bought 33 laying hens, so now we have just 50 hens. Is that right?
5. Uncle Frank had 85 ducks. He sold 12 at the market. Then one day 5 of those he had left got out and swam down the creek, and never came back. How many ducks did Uncle Frank then have left?
6. Irene picked 21 boxes of currants. She made jelly out of 10 boxes and sold the rest at 5¢ a box. How much did she get for those she sold?
7. John bought 40 marbles. He lost 6. He gave half of those left to Jane, so they each have ...... marbles.
8. Father drove our auto 25 miles an hour for 4 hours. Then Mother drove for him for 18 miles, which brought us right to Uncle Will's farm. How far is it to Uncle Will's farm?
9. Jean is going to spend 3 weeks at Aunt Laura's. Jack is going, too, for 10 days, and after that he is going to Scout Camp. How long will Jean be at Aunt Laura's without Jack?
10. Henry planted 6 rows of beans. He picked 16 quarts of beans from each row. He sold them for 6¢ a quart. How much did he get for his beans?
11. Tom, Joe, and Bill together picked 18 quarts of wild blackberries. They sold them for 17¢ a quart. They divided the money equally. How much did each one get?
12. There are 9 girls in my club, but some are moving away, so we have

decided to divide our $2.25 equally. I have saved 75¢ and with the money I receive from the club I shall have $...... in all.

13. Dick had 65¢ to buy favors for his party. He saw puzzles for 8¢ each. He bought all he could. When he got his change, he thought "If I had 4¢ more I could also buy a ......-cent top."
14. A grocer bought 6 big oranges for 24¢. He sold them for 8¢ apiece. "I have gained ...... cents," said the grocer.
15. It costs $8.00 a week for board at the Scout Camp. If John stays at the camp for 14 days, how much in all does he pay for board?
16. Father bought 1000 young trees. He sold ½ of them to Uncle George. Father planted ½ of his and gave the rest to my Scout Troop. How many trees did the Scout Troop get?
17. At 5 minutes after 12 Father called up and said, "I'll be home for lunch in 10 minutes." But it was half-past 12 when he came, so he was ...... minutes later than he thought he would be.
18. The box in which James keeps the oats for his pony holds 15 quarts. He feeds his pony 3 pints of oats a day, so he has to fill his box every ...... days.
19. It is 3 miles from our house to Father's office. He drives there and back every day but Sunday, so in a week he drives ...... miles.
20. Mrs. Green made a gallon of ice cream for our class sale. She sold the ice cream for 5¢ a dish. There were 8 dishes to each quart. How much money did she give the teacher?
21. Ira said, "I need 4 yards of wire for my round flower bed, and 13 feet for the long bed by the porch." Wire comes in 25-foot and 50-foot rolls. Will the small roll be enough?
22. Ellen has a book with 12 chapters in it. She read up to Chapter V last night. She read 2 more chapters to-night. How many more chapters has she to read?
23. Jean had $14.50 in the bank. She was given $5.00 for Christmas, so she bought a pair of skates for $2.75 out of it and put the rest in the bank. Now she has $...... in the bank.
24. Our baby weighed 8 pounds and 2 ounces when it was born. If it gains 3 ounces every week, at the end of 4 weeks it will weigh ...... pounds and ...... ounces.
25. Mrs. Wilson bought a yard of linen lace to trim her towel. She used 17 inches of lace on each end of the towel, and gave me the rest. How many inches of lace did she give me?

Grade 4

Three-Step

1. Aunt Carrie gave me a dime and a nickel. She said, "Please buy me 7 two-cent stamps. You may keep the change." How much did I get?
2. Joe's father hauls sand and stone. He has 2 trucks with 4 wheels each and 2 trucks with 6 wheels each. If he needed new tires for all his trucks, how many would he have to buy?
3. On Mother's birthday we went to the Coffee Shop for supper and I

counted 6 tables with 4 chairs each and 6 tables with 2 chairs each. How many people could sit down to eat at the same time?

4. Uncle Jim's cow gives 4 quarts of milk in the morning and the same amount at night. He said, "I wouldn't sell that cow. She gives ...... gallons of milk a week."
5. Uncle Bob said to Betty, "Here are 2 silver dollars and 6 one-dollar bills. If you can tell me how many dimes there are in these pieces, I shall give you half of the right number of dimes." Betty told Uncle Bob the right answer, and he gave her ...... dimes.
6. Mrs. Jackson bought 3 yards of lace. She used 40 inches on each of 2 handkerchiefs. Ann asked for the rest to trim a doll's apron. How much did she get?
7. Last week the milk man brought us 7 quarts of milk at 16¢ a quart and 2 bottles of cream at 27¢ a bottle. How much was our bill?
8. Roddy went to the hardware store to buy some nails. They cost 24¢. He saw a dog collar for 15¢ and bought it. He gave the man 50¢. He spent all his change for a 5-cent package of gum and a ride home on the trolley. How much was the trolley fare?
9. Nick had 90¢ and Nancy had 80¢. Nick said, "I'll take ⅓ of my money if you will take ¼ of yours, and together we can buy Father a birthday present which will cost ......¢.
10. The 9 boys in our team decided to get baseball suits. We found if we each bought separately, the cost was $4.50 apiece, but if we ordered 9 at once the cost was $3.98 each. So we ordered together and saved $...... on all the suits.
11. Alice bought 11 greeting cards at 3¢ each. She sent them in the mail and put a 2-cent stamp on each envelope. How much did Alice spend in all?
12. The florist had tulip bulbs to sell for 3¢ each. Mabel had 48¢ and Joan had 57¢. Each spent all their money for bulbs. How many more bulbs did Joan get than Mabel?
13. In our school there are 3 rooms upstairs and 4 rooms downstairs. In each room there are 7 rows of seats. Six children sit in each row, so there are ...... children in our school.
14. Eight Scouts on a hike bought 24 sandwiches at 5¢ each, and 32¢ worth of oranges. They shared the cost equally, so how many cents did each one pay?
15. Ralph raised 9 bushels of potatoes in his garden. He sold 4 bushels at 80¢ each and 5 bushels at 90¢ each. How much money did he get for all his potatoes?
16. Joe bought 200 papers to-day for $2.75 and sold them all for 2¢ apiece. He gave ⅕ of what he had made to his mother. Joe said, "Here is ......¢ for you, Mother."
17. Rita was telling her mother about the parade. She said, "There were 42 Boy Scouts marching 6 abreast and 36 Girl Scouts marching 6 abreast." How many rows of Scouts were there?
18. White clover seed is 59¢ a pound and grass seed is 42¢ a pound.

Father said, "Go get 2 pounds of each and take the right change with you." How much money did I take along?

19. There are 8 houses in the block where Peggy lives. In each of 6 of the houses 2 children live. Three children live in each of the other 2 houses, so there are ...... children in the block where Peggy lives.
20. Dick went to the store and bought ½ dozen eggs at 56¢ a dozen and ¼ pound of cheese at 48¢ a pound. How much was Dick's bill?
21. Mr. Lewis had $794 in the bank. He drew out $145 to pay for a new radio. Then he sold a cow for $180 and put the money in the bank. Next day his brother asked him for $125. When he took that from the bank, how much had he left?
22. Father had a piece of rope 50 feet long. He cut off 5 yards for a tow rope and Bobby took 4 feet to tie his goat with. How many feet of rope were left?
23. There are 12 girls in our Glee Club. We had $3.00 in our club treasury. We bought a song book for each 2 girls. We paid 45¢ for each book. After the books were paid for, how much did we have left?
24. We have 3 rooms downstairs with 3 windows in each room. Each window has 4 panes in it. We have one door with 15 panes in it. When Lucy finished washing all the windows and the door, she said, "I've washed ...... panes of glass."
25. I watched our grocer unpack cans of tomatoes. He had 3 boxes of cans. There were 2 layers in each box. In each layer there were 3 rows of cans with 4 cans in a row, so he unpacked ...... cans of tomatoes.

GRADE 4

Four-Step

1. James hung out 3 wren's houses this spring. A pair of wrens nested in each one. Each mother wren laid 4 eggs. All the eggs hatched but three. How many wrens, young and old, lived in James's houses?
2. There are 8 houses in our block. In each of 3 of them 4 children live. In each of the other houses 3 children live. So in our block there are ...... children.
3. Eight Scouts on a hike bought 24 sandwiches at 5¢ each and 2 dozen oranges at 36¢ a dozen. They shared the cost equally. How much did each have to pay?
4. Louise went to the store and bought ¼ of a pound of cheese at 48¢ a pound and a can of tomatoes for 28¢. She gave the man a quarter, a nickel and a ....
5. Last week Fred mowed 3 lawns and got 25¢ for mowing each one. This week he weeded in the garden for 4 hours and got 20¢ an hour. He went to the fair and spent all the money he had earned but a quarter. How much did Fred spend at the fair?
6. Tom and Donald saw marbles for sale in a window; they were 3 for 1¢. Tom said, "I have an idea. If you will spend ⅓ of the 15¢ which your mother gave you, I will spend ¼ of the 40¢ which I have in my pocket and together we can buy ...... marbles."

7. Miss Burns takes the money we pay for food in our school lunch room. I watched her count the money this noon. She had 56 pennies, 24 nickels, 17 dimes, and 5 quarters. How much money was that in all?
8. Edgar paid $3.75 for a pair of shoes, 50¢ a pair for 2 pairs of socks, 75¢ apiece for 2 ties, and 10¢ apiece for 4 handkerchiefs. How much did Edgar's new clothes cost?
9. Alan wants to buy a camera that costs $4.54. He already had $1.60 in his bank. This week he mowed 2 lawns and got 25¢ for each one. He worked 6 hours in the garden at 15¢ an hour. How much more does he still need to earn in order to buy the camera?
10. A farmer raised 240 bushels of potatoes on one field and 205 bushels on another. He kept 20 bushels for seed and 18 bushels to eat. He sold the rest for 85¢ a bushel. How much did he get for his potatoes?
11. Lucy had $12.15 in her purse when she went downtown. She bought 3 handkerchiefs at 15¢ each, a pair of slippers for $3.75, and ¼ yards of ribbon for her watch at 40¢ a yard. How much money had she left in her purse?
12. Howard had 7 half-dollars and 3 dimes. He spent $1.65 for a knife with 3 blades. How much money did he have left?
13. A farmer picked 204 bushels of apples in one orchard and 80 bushels in another. He kept 24 bushels for his own use. He sold the rest for 60¢ a bushel. He put half his money in the bank. How much did he put in the bank?
14. The boys in our baseball team bought 2 bats at 39¢ each, a mask for $1.49, 2 mitts at 99¢ each, and 3 balls at 59¢ each. How much did they spend in all?
15. Julia bought ¼ of a pound of butter at 52¢ a pound, 2 loaves of bread at 8¢ each, ½ pound of cottage cheese at 26¢ a pound. She made picnic sandwiches and used up all the things she had bought, so the sandwiches cost her ......¢.
16. Father's garden is 25 feet wide and 40 feet long. He bought 150 feet of wire. He took enough to fence his garden all around and gave the rest to me, so I got ...... feet of wire.
17. Carl earned 15¢ raking leaves. He beat the rugs twice and got 10¢ each time. He spent all this money for pencils that cost 5¢ each. He gave 2 pencils to his father. How many pencils did Carl have left for himself?
18. Louise had 6 weeks' vacation. She spent 6 days of it sewing and 8 days on an auto trip. Then she spent one week at Grandmother's and the rest of the time at the seashore. How many weeks did she spend at the seashore?
19. Father paid 40¢ a pound for a chicken that weighed 4 pounds, 6¢ a pound for 2 pounds of onions, and 80¢ a pound for ½ pound of marshmallows. He said, "I've spent $...... toward the Sunday dinner."
20. Lester gathered 2 dozen eggs on Monday and broke 6 eggs as he stumbled on his way to the house. He gathered 27 eggs on Tuesday,

25 eggs on Wednesday, 28 eggs on Thursday, and 16 eggs on Friday without breaking a single one. On Friday night Lester said, "I have packed ...... dozen unbroken eggs so far this week."

21. Uncle George has 2 horses. He feeds each one 4 quarts of oats three times a day. Cousin Walter feeds his pony 2 quarts of oats twice a day. How many quarts of oats are fed to the horses and the pony in a day?
22. Uncle George had 2 bushels of seed corn. He sowed 2 pecks in one field and 3 pecks in another. He said, "I am going to sell what I have left at 85¢ a peck." How much money did Uncle George get?
23. The suits for the 9 girls in my swimming class cost $2.00 each, and the caps cost $4.05. Because we bought them all at once, the store took off $2.25 from the bill, so each girl had to pay only $...... in all.
24. We had 2 lambs that had lost their mothers. From a bottle we fed them each 3 pints of milk a day. I thought, "In a week those rascals drink ...... gallons of milk and 1 quart more."
25. Father bought an 8-pound turkey for Christmas. It cost 65¢ a pound. He bought 2 quarts of cranberries at 14¢ a quart. What was his change from $6.00?

ARITHMETIC REASONING PROBLEMS

Grades 5, 6, and 7

One-Step

1. To-day Mother bought me a pair of socks for 25¢, a dress for $2.36, a coat for $8.94, and a pair of shoes for $3.50. What did all my new clothes cost?
2. Alan had a rope 15 feet long. He cut off 6 feet to tie on his sled. How long was the piece that was left?
3. On Saturday Father took us all to the circus. He bought 5 tickets and paid $1.05 for each one. How much did he pay in all for the tickets?
4. Last summer there was no rain for 42 days. Mr. Farmer said, "We've had no rain for ...... weeks."
5. Father bought a suit for $43.50 and an overcoat for $37.25. How much did he pay the man for both of them?
6. Uncle Fred sent me a baby alligator from Florida. It was 9 inches long when it came and now it is 17 inches long. How many inches has it grown since I received it?
7. Aunt Fanny bought skates for each of her 3 nephews for Christmas. She paid $2.60 a pair for the skates. How much did the skates cost her in all?
8. During vacation Walter worked for $8 a week and earned $72. How many weeks did he work?
9. In our school there are 76 children in the first grade, 69 in the second, 64 in the third, and 45 in the fourth grade. How many children are there in the four grades?
10. Father was looking at bathrobes marked $13.50. He said, "I like this one, but it is more than I can pay." The clerk said, "If this one

fits you, I'll take off $2.25." The robe fitted father. How much did Father pay for it?

11. We pay $65 a month rent for our house. How much does it cost us to live in our house for a year?
12. It is 323 miles from our house to our summer camp. It took us 9 hours to drive there. How many miles an hour did we drive?
13. The county fair ran for five days. The first day 486 persons bought tickets, the next day 519, the third day 737, the fourth day 683, and the last day 960. So ...... persons in all attended the fair.
14. Uncle Roger bought a new car which cost $1250. He turned in his old car and was allowed $325 for it, so he had to pay the salesman only $...
15. Aunt Susan makes delicious jelly and sells it in the tea room that she runs in summer. This year she sold 807 glasses of jelly at 45¢ a glass. How much did she get in all for it?
16. Nine of us Girl Scouts spent the week-end in camp and spent $4.68 for supplies. We shared the cost equally, so my share was ......¢.
17. Mother bought a bed for $4.65, a mattress for $5.15, a pillow for $3.09 and a mirror for 69¢ for my room in our summer camp. How much did the furnishings for my room cost altogether?
18. Cousin Louis spent $14.30 in planting his potatoes. He got $71.25 for them when he sold them. How much profit did he make on his potatoes?
19. There are 12 windows in our sun-porch. Ida washed one and found it took her 8 minutes. She said, "I ought to be able to do all these windows in ...... minutes."
20. Mother lacked ⅛ of a yard of gingham to finish my dress. I went to the store for it. The gingham cost 40¢ a yard. How much did I pay for what I bought?
21. Cousin Paul lives on a farm and has so far to go to school that he drives a pony. The pony cost $85, the cart $34.50, and the harness $13.95. How much did the outfit cost in all?
22. Father earned $335.00 last month and spent $208.60. How much did he have left?
23. The painter thinks it will take 7 gallons of paint to cover our house. At $3.25 a gallon it will cost $...... for the paint for our house.
24. It is 108 miles around Orchard Lake. A steamer goes around the lake every day. It goes 18 miles an hour. How many hours will it take to make the trip?
25. Uncle Charles paid $36.25 for a saddle, $3.25 for a saddle blanket, $3.80 for a bridle, and $3.68 for spurs. He said, "This riding outfit cost me $.......

Grades 5, 6, and 7

Two-Step

1. Albert's harness for his goat cost $4.55 and his cart cost $5.00. He had saved $12.00 so he had $...... left after paying for the harness and the cart.
2. Father had a 25-foot roll of garden wire. He cut off 4 yards and gave

it to Alice to put around her tulip bed. How many feet of wire did he have left?

3. It was 15 minutes after the hour when the telephone rang. Father said, "I will be home in 20 minutes." But it was 50 minutes after the hour when he arrived. How late was he?
4. It took 3½ yards of cloth for my dress. I paid $1.48 a yard for the cloth. Mrs. Taylor charged $3.50 to make my dress, so in all it cost $.......
5. Daddy took Mother to the theater. He paid $1.65 apiece for the tickets and he paid 35¢ for the music of the song they liked best. How much did Daddy spend in all?
6. Herbert bought a tennis racket for $2.98 and 3 tennis balls at 39¢ each. How much did he spend in all?
7. Cousin Robert says the engine in his boat uses up 2 quarts of gasoline an hour when running full power. How far can the boat run on 5 gallons of gasoline?
8. Alice was given $5.00 for a birthday present. She put half of it in her dime bank and spent $2.00 for a bracelet. She spent the remaining ......¢ for candy.
9. David bought ¾ of a pound of cheese at 40 cents a pound and a box of crackers for 18¢. How much did he pay for both?
10. Mr. Mann gathered 384 eggs last week. Mrs. Mann used a dozen for cooking. How many dozen did Mr. Mann have left to sell?
11. Polly earned $1.70. She bought a package of flower seed for 5¢, and with the rest of her money bought rose bushes at 33¢ each. Do you know how many rose bushes she bought?
12. Peter weeded in the garden 2 hours a day for 4 days. Father paid him 15¢ an hour. How much did he earn weeding?
13. My pet alligator was 9 inches long when I got it. Now it is 2 feet long, so it must have grown ...... inches.
14. Uncle Roger bought each of his 3 nephews a Boy Scout knife for $1.48 each, and a knife for himself for $1.79. How much did he spend for knives?
15. During vacation Sara worked for 6 weeks at $8 a week. She put ⅔ of her money in the bank, so she saved $.......
16. Uncle Roger had $5,000 in the bank when he bought his new car. The price of the new car was $1,250. He turned in his old car and was allowed $325 for it. How much money had he left in the bank, after he had paid for the car?
17. Just before Christmas Aunt Fanny bought 3 pairs of skates at $2.60 a pair. She gave the man a ten-dollar bill. How much money should she get back?
18. Father bought a farm for $7,850. He spent $1,069 for repairs on the house and barn. Then he sold it for $10,000. How much did he gain?
19. My brother and 4 of his friends bought a boat for $225.00 and a motor for it for $136.95. The five of them shared the cost equally. How much was my brother's share?

20. Edwin's garden is 15 feet long and 12 feet wide. He spaded up ⅓ of it this week. He said, "I've spaded up ...... square feet in my garden."
21. At the beginning of last month my big brother had $124.60 in the bank. He put in $237.50 at the middle of the month, and at the end of the month drew out $189.75. How much money did he have in the bank at the beginning of this month?
22. I saw marbles in the toy shop marked "3 for 1¢." I bought 36 of them and an airplane kite for 27¢. How much did I spend in all?
23. Father's wagon weighed 1100 pounds. Filled with wheat the wagon weighed 3920 pounds. A bushel of wheat weighs 60 pounds. Father said, "There must be ...... bushels of wheat in that load."
24. Helen bought 4⅛ yards of gingham at 48¢ a yard and a patent leather belt for 29¢. How much did she spend?
25. Mother had 8 yards of cloth. She used 3¼ yards for a dress for Ann, 2¾ yards for a dress for Mary, and the rest for Bobby's rompers. How much was left for rompers?

Grades 5, 6, and 7

Three-Step

1. Henry had $14.80 in the bank when he bought his new bicycle. The price of it was $10.00. The man said, "If you will pay me cash, I will make the price $3 less." Henry said, "I will pay cash, but I also need a bicycle pump." The man said, "A new bicycle pump will cost you 50¢ more." How much money was left in Henry's bank after he had paid for his bicycle and a new bicycle pump?
2. Polly bought 3 rose bushes at 33¢ each and 3 vines at $1.98 each. How much did she pay out in all?
3. Nick put 19 cents into the school savings bank each week for 32 weeks, and Nancy put 17 cents into the bank each week for 30 weeks. How much more did Nick save than Nancy?
4. During vacation Sally worked for 8 weeks for $9 a week. She gave ⅙ of her money to her mother and spent the rest for new school clothes. How much did Sally spend?
5. Aunt Susan served 1614 meals at 50¢ each in her tea room this summer, and sold 807 glasses of jelly at 45¢ a glass, so she must have taken in $...... altogether.
6. Uncle Roger had $5,000 in the bank when he bought his new car. The price of the new car was $1,250. The cost of the car license was $12.50. He turned in his old car and was allowed $325 for it. How much money had he left in the bank, after he had paid for his car and his car license?
7. Our church held a two-day fair. There was a total of 368 people who attended during these days. ¼ of these people were children who were admitted free of charge. The others had to pay 55¢ each, so we took in $...... at the door.
8. Five boys were to share equally the cost of their new boat equipment. They had bought a pair of oars at $4.37 a pair, 5 life-saver cushions

at $1.75 each, and a fire extinguisher at $1.88. What was each one's share?

9. The ladies in Mother's club bought 208 rose bushes at 46¢ each to plant along our new park drive. There are 60 ladies in the club and they each paid $2.00 toward the rose fund. After they had paid for the rose bushes, how much did they have left to pay toward planting them?
10. Edwin's garden is 15 feet long and 12 feet wide. Elsie's garden is 15 feet long and 8 feet wide. Edwin said, "I have ...... square feet more of garden to weed than you have, Elsie."
11. Uncle George raised 1683 bushels of potatoes on one field of 9 acres, and 1624 bushels of potatoes on another field of 8 acres. He said, "My 8-acre field yielded ...... more bushels per acre than my 9-acre field."
12. Aunt Ellen pays $55 a month rent for her house. A teacher rents a room of Aunt Ellen for 9 months of the year at $12 a month. Using the teacher's room rent toward paying her rent, how much does Aunt Ellen pay out of her own money to rent her house for a year?
13. Our grocer bought a box containing 8 dozen apples for $3.20. He sold the apples for 5¢ each. How much did he gain on that box of apples?
14. Uncle Charles' strawberry patch yielded 37 crates of berries last week. There are 32 quarts of berries in a crate. He sold the berries for 18¢ a quart. After selling the berries, Uncle Charles took the money and bought some new crates which cost him $13.12. How much money did he have left to put in the bank?
15. I picked strawberries for Uncle Charles last week. One day I picked 3 crates holding 32 quarts of berries each. Another day I picked 64 quarts which Mary canned for the house. Uncle Charles paid me 3¢ a quart for my work. How much money did I receive in all?
16. The pupils in our school held a Flower Show. The tickets for parents were 25¢ each and the tickets for children were 10¢ each. In all we sold $8.10 worth of tickets. 18 of these tickets were sold at 25¢ each. How many tickets were sold at 10¢ each, in order to make our $8.10?
17. Last month Walter earned $18 and saved 1/9 of it; this month he earned $21 and saved 1/7 of it. He said, "I was able to save $...... more this month than last month."
18. One day Walter found $16. He spent 1/2 of it for a raincoat and 1/4 of it for shoes. How much money did he spend?
19. This month Walter spent $18. He spent 1/6 of it for a mirror for Mother's birthday present. The rest he spent for radio supplies, paying $10.00 for a battery and the rest for tubes. How much did the tubes cost?
20. Father bought 6 1/2 tons of coal at $10 a ton and 3 1/4 tons at $12 a ton. Father paid $...... in all, and said, "That will keep us warm for a while."
21. Grace brought home 1/8 of a pound of candied fruit that cost 80¢ a

pound and ½ a pound of chocolates at 40¢ a pound. How much did she pay for the treat?

22. Uncle Jim bought 8 gallons of gasoline at 18¢ a gallon and 1 quart of oil at 20¢ a quart. He found he had only a ten-dollar bill. How much change did he get?
23. Mother is buying a new sewing machine. It costs $49. She paid $4 down in the beginning. She pays $5 each month. She has paid for 4 months. She said, "I still have $...... to pay on that machine."
24. I went with Father on a trip to the city. He let me buy the railroad tickets. His ticket was $5.50 and mine ½ as much as his. I gave the man $10.00. How much money was I to get back from the man?
25. As I was helping Uncle Henry drive his sheep to the new pasture, he said, "I paid $9 apiece for these sheep. They cost me $1503. I think I can sell them for $12 apiece. If I do, I'll make $...... on them."

Grades 5, 6, and 7

Four-Step

1. Tom's hens laid 18 eggs each for 9 days. He sold these eggs for 49¢ a dozen. Half the money he got for eggs he spent for chicken feed. How much did Tom spend for chicken feed?
2. Uncle Roger had $5000 in the bank when he bought his new car. The price of it was $1250. The dealer said, "Because you can pay cash, I will make the price of the car $50 less." Uncle Roger paid cash for the car. He turned in his old car and was allowed $325 for it. After he had bought the new car, he had to spend $56 to insure it. How much money did Uncle Roger have left in the bank after he had insured the new car?
3. Jim bought 8 gallons of gasoline at 18¢ a gallon and 2½ quarts of oil at 20¢ a quart. What was his change from a 5-dollar bill?
4. Mr. Farmer had 40 crates of strawberries yesterday. There are 32 quarts in a crate and he paid 2¢ a quart to have them picked. He paid 10¢ a crate to have them hauled to market. How many dollars did Mr. Farmer have to pay to have the berries picked and hauled?
5. Uncle Frank sold 4 ducks at $1.60 each and 4 turkeys at $4.90 each. With his money he bought seed corn at $2 a bushel. How many bushels of seed corn did Uncle Frank buy?
6. Our fourth-grade class gave our teacher a basket of fruit at Thanksgiving time. We paid 45¢ for the basket. We put in ½ dozen oranges at 48¢ a dozen, a package of dates at 15¢, 6 red apples at 2 for 5¢, and a 10-cent grapefruit. We tied a green ribbon that cost 10¢ on the handle. The present cost us ......¢.
7. We are planning to plant a hedge all round our back lawn which is 50 feet long and 25 feet wide. If we plant 2 hedge bushes to each foot we shall need ...... bushes.
8. To ride in a taxicab costs 15¢ for the first ¼ mile. It costs 5¢ for each ¼ mile after the first. How much would it cost you to ride 2 miles?

9. Mr. Hill earns $1,200 a year. He spends $25 a month for rent, and $10 a week for other expenses, so he has $...... a year left that he can save.
10. This summer Cousin Jim sold books. He made $10 the first week. Each week after that he made $2 more than the week before. How many dollars did Cousin Jim make the first 4 weeks?
11. Paul counted 660 steps he took in going to the post office. He measured his step and found it was 24 inches long. He figures that he walked half a mile in going to the post office and back. Was that right? (There are 5,280 feet in a mile.)
12. Phil was wondering how many hours of recess he had had during this year. He thought, "I have 15 minutes in the morning and 15 minutes in the afternoon, for five days a week, and 32 weeks. That makes ...... hours."
13. Our grocer bought a box of soap that had 6 layers of soap in it. Each layer had 6 rows of soap with 6 bars of soap in a row. He sold the soap at 2 bars for 15¢. He said, "I get $...... for a box of this soap."
14. Mabel bought ¼ of a pound of butter at 60¢ a pound and ¾ of a dozen eggs at 56¢ a dozen. She gave a dollar bill to the man and brought home ......¢ change.
15. When we started on our trip our gasoline tank was full, and when we got back, it was ¼ full. Our tank holds 12 gallons of gasoline. When we started, our speedometer read 15,621 miles; when we got back it read 15,747 miles. How many miles did we go on each gallon of gasoline?
16. Mildred is the cashier at the big high school cafeteria. This noon when she counted the bills she had taken in, she had 2 ten-dollar bills, 6 five-dollar bills, 4 two-dollar bills, and 12 one-dollar bills. How many dollars did she have in all?
17. Mary has a garden 15 feet long and 8 feet wide. She plans to plant 2 bulbs in each square foot. The bulbs cost 49¢ a dozen, so enough to plant her garden will cost $.......
18. Laura bought 3 yards of gingham for aprons at 45¢ a yard, and 3 yards of velvet ribbon at 65¢ a yard. What was Laura's change from a five-dollar bill?
19. Our school gave an entertainment. There were 426 tickets sold. One-half the tickets were bought by the children at 10¢ each. The rest of the tickets were bought by the parents at 25¢ each. How much money was taken in at our entertainment?
20. William had 7 half-dollars and 3 quarters. He spent $1.79 for a pair of tennis shoes. How much money did William have left?
21. A farmer I know sold 48 hogs for $37.60 each. He put $1,000 of this money in the bank. He spent ¼ of the money that was left for a radio, and gave the rest to his wife as a present. How much did his wife receive?
22. As Edwin was mowing the lawn, he thought, "This lot is 15 yards wide and 45 yards long. The house is 10 yards wide and 12 yards long

and the sidewalk takes up 15 square yards. I wonder how many square yards of lawn that leaves." Can you tell him?

23. Ernest got a job picking cherries for 3¢ a quart. He said to his younger brother, Don, "If you will help me, I will give you one-third of all the money that we both earn." It happened that Ernest picked 4 gallons of cherries and Don picked 6 gallons. There are 4 quarts in a gallon. How much did Don get from his brother Ernest?
24. We paid $2.45 a square yard to have our driveway made and $100.00 extra to have the curbing made. The driveway is 6 feet wide and 81 feet long, so it cost us $...... in all.
25. Mary bought ½ pound of nuts at 38¢ a pound, ¼ pound of butter at 60¢ a pound, 5 pounds of sugar at 7¢ a pound, and a can of milk for 15¢. How much did all these articles cost?

CHAPTER V

STATISTICAL RESULTS

The possible practice effect which might result in an experiment of this kind should always be considered when noting or interpreting further the statistical data which follow.

Preliminary and Final Tests

Table 3 shows the results of the preliminary and the final tests, with the average gains made between them, from Grades 3, 4, 5, 6, and 7.

It will be seen that the results in both preliminary and final tests show a very consistent decrease from the upper to the lower groups in ability to solve problems. There seems to be a very consistent decrease also in ability to solve verbal problems correctly as pupils go from one-step problems to four-step problems.

The gains which were made in the final tests, after drill had been given on computational difficulties, are practically identical for all steps and all groups.

It should be noted that if we expect better than a 60 per cent mastery or achievement for an entire grade in problem-solving work, after drill has been given on computational difficulties, that Grades 3, 4, and 5 achieved this on two-step problems. In Grades 4 and 5, the pupils do three-step problems. In the sixth and seventh grades, where the fifth grade problems were administered, we find approximately 60 per cent achievement or better on one-, two-, three-, and four-step problems for all groups.[1] The lower group in the sixth grade apparently did not do so well as the lower group in the fifth grade. The upper group in the seventh grade was able to solve four-step problems with 68 per cent mastery or achievement.

Since there is a very high agreement between the data in Table 3 and the results shown on the inventory tests, it is evident

[1] See Appendix for modification of last three statements.

TABLE 3

RESULTS OF PRELIMINARY AND FINAL TESTS BEFORE AND AFTER REMEDIAL INSTRUCTION, INCLUDING GAINS MADE, BY GROUPS, AND FOR GRADES 3, 4, 5, 6, AND 7

GROUP		PRELIM. TEST				FINAL TEST				GAINS MADE			
		Number of Steps in Problem Tests											
		1	2	3	4	1	2	3	4	1	2	3	4
GRADE 3													
Upper group (86 pupils)	Av. No. problems solved correctly per pupil	13	14			19	18			6	4		
	Per cent problems solved correctly per pupil	52	56			76	72						
Middle group (137 pupils)	Av. No. problems solved correctly per pupil	9	10			16	14			6	4		
	Per cent problems solved correctly per pupil	36	40			64	56						
Lower group (51 pupils)	Av. No. problems solved correctly per pupil	9	10			15	13			6	3		
	Per cent problems solved correctly per pupil	36	40			60	52						
Entire grade (274 pupils)	Av. No. problems solved correctly per pupil	10	11			17	15			7	4		
	Per cent problems solved correctly per pupil	40	44			68	60						
GRADE 4													
Upper group (94 pupils)	Av. No. problems solved correctly per pupil	18	9	8	6	22	14	12	10	4	5	4	4
	Per cent problems solved correctly per pupil	72	36	32	24	88	56	48	40				
Middle group (122 pupils)	Av. No. problems solved correctly per pupil	15	7	5	4	19	11	8	6	4	4	3	2
	Per cent problems solved correctly per pupil	60	28	20	16	76	44	32	24				
Lower group (54 pupils)	Av. No. problems solved correctly per pupil	14	9	5	3	18	10	8	6	4	3	3	2
	Per cent problems solved correctly per pupil	56	36	20	16	72	40	32	24				
Entire grade (270 pupils)	Av. No. problems solved correctly per pupil	16	8	6	4	20	12	10	7	4	4	3	3
	Per cent problems solved correctly per pupil	64	32	24	16	80	48	40	28				
GRADE 5													
Upper group (60 pupils)	Av. No. problems solved correctly per pupil	18	11	9	6	19	13	12	10	4	2	4	3
	Per cent problems solved correctly per pupil	72	44	36	24	76	52	48	40				
Middle group (141 pupils)	Av. No. problems solved correctly per pupil	16	9	6	4	19	12	10	8	3	3	4	3
	Per cent problems solved correctly per pupil	64	36	24	16	76	48	40	32				
Lower group	Av. No. problems solved correctly per pupil	14	8	6	4	19	10	9	6	5	2	3	2
	Per cent problems solved correctly per pupil	56	32	24	16	76	40	36	24				
Entire grade (270 pupils)	Av. No. problems solved correctly per pupil	16	9	7	4	19	12	10	8	3	3	4	3
	Per cent problems solved correctly per pupil	64	36	28	16	76	48	40	32				

TABLE 3—(*Continued*)

Group		Prelim. Test				Final Test				Gain Made			
		Number of Steps in Problem Tests											
		1	2	3	4	1	2	3	4	1	2	3	4
GRADE 6													
Upper group (55 pupils)	Av. No. problems solved correctly per pupil	21	16	13	10	23	20	18	14	2	4	5	4
	Per cent problems solved correctly per pupil	84	64	52	40	92	80	72	56				
Middle group (252 pupils)	Av. No. problems solved correctly per pupil	19	12	9	6	22	17	15	11	3	5	6	5
	Per cent problems solved correctly per pupil	76	48	36	24	88	68	60	44				
Lower group (87 pupils)	Av. No. problems solved correctly per pupil	17	9	7	5	12	13	12	9	3	4	5	4
	Per cent problems solved correctly per pupil	68	36	28	20	48	52	48	36				
Entire grade (394 pupils)	Av. No. problems solved correctly per pupil	19	12	9	6	20	16	14	11	3	5	5	5
	Per cent problems solved correctly per pupil	76	48	36	24	80	64	56	44				
GRADE 7													
Upper group (67 pupils)	Av. No. problems solved correctly per pupil	22	19	18	12	23	21	20	17	3	2	2	3
	Per cent problems solved correctly per pupil	88	76	72	48	92	84	80	68				
Middle group (192 pupils)	Av. No. problems solved correctly per pupil	21	11	14	10	22	19	17	14	9	3	3	3
	Per cent problems solved correctly per pupil	84	44	56	40	88	64	68	56				
Lower group (57 pupils)	Av. No. problems solved correctly per pupil	19	13	11	8	21	17	14	12	2	3	3	4
	Per cent problems solved correctly per pupil	76	52	44	32	84	68	56	48				
Entire grade (316 pupils)	Av. No. problems solved correctly per pupil	21	16	14	11	22	17	17	14	9	3	3	3
	Per cent problems solved correctly per pupil	84	64	56	44	88	68	68	56				

Table 3 is read as follows:—For Grade 3 there are eighty-six pupils in the upper group having I.Q.s of 116 and upward; 137 pupils in the middle group with I.Q.s ranging from 100 to 115, inclusive; and 51 pupils in the lower group with I.Q. scores of 99 or less. In the upper group the average number of one-step problems per pupil solved correctly in the preliminary test was 13; in the final test, 19; this made a gain of six problems on one-step problems, after drill had been given upon computational weaknesses. This also means that of the one-step problems, 48 per cent of them per pupil were not solved correctly in the preliminary test and 24 per cent of them per pupil were not solved correctly in the final test. Of the two-step problems, the upper group only solved an average of 14 per pupil correctly in the preliminary test; and 18 per pupil correctly in the final test. This means that, on the average, 44 per cent of the two-step problems were not solved correctly in the preliminary test; and 28 per cent per pupil were not solved correctly in the final test. Similar details are to be interpreted in like manner for each group in all grades, and for the combined groups of all grades tested.

that problem-solving ability is greatly affected by computational ability. This bears out Washburne's findings.[2]

There also seems to be some agreement between these results and those found in "errors due entirely to erratic performance." This may indicate that problem-solving results are influenced by both computational and erratic performance.

Inventory Tests

Table 4 sets forth the results of the inventory test for Grades 3, 4, 5, 6, and 7. An examination of this table shows that the number of complete examples solved correctly on the inventory test decreased for all groups in each grade from one-step problems to four-step problems. This means, of course, that the computational errors must have increased with the number of steps involved in the problem, since none of the three- or four-step problems were counted right unless *all* the arithmetical operations or processes involved in any one problem were correct. This is true.

The per cent of computational examples not solved correctly increases quite consistently from one-step examples to four-step examples for all grades and all groups. The per cent of increase is especially high for Grade 4. For Grades 6 and 7 the increase becomes lower in direct ratio to the ascending grades. This would point to varying degrees of computational difficulty and computational mastery.

The large and consistent increase in the per cent of examples not solved correctly, as pupils go from one-step examples to four-step examples, points to a high degree of erratic performance or unstable computational ability, since the arithmetical operations in three- and four-step problems were no more difficult than those found in one- and two-step problems.

By a comparison of the per cent of problems not solved correctly in the inventory test with those not solved correctly in the preliminary tests (Table 3), we find corresponding increases. This situation, combined with the fact that Table 6 shows a high per cent of errors that are due to erratic performance, would indicate that unstable computational ability figures rather formidably in "problem solving."

It should be noted also that although the results of the inven-

[2] Washburne, C. W., *Teacher's Manual, Individual Arithmetic*, p. 12.

TABLE 4

RESULTS OF INVENTORY TEST, BY GROUPS, FOR GRADES 3, 4, 5, 6, and 7

GROUP		PROBLEM TESTS			
		One-Step	Two-Step	Three-Step	Four-Step
GRADE 3					
Upper group (86 pupils)	Av. No. examples solved correctly per pupil...... Per cent examples NOT solved correctly, per pupil	15 40	13 48		
Middle group (137 pupils)	Av. No. examples solved correctly per pupil...... Per cent examples NOT solved correctly, per pupil	14 44	11 56		
Lower group (51 pupils)	Av. No. examples solved correctly per pupil...... Per cent examples NOT solved correctly, per pupil	15 60	11 40		
Entire grade (274 pupils)	Av. No. examples solved correctly per pupil...... Per cent examples NOT solved correctly, per pupil	15 60	11 56		
GRADE 4					
Upper group (94 pupils)	Av. No. examples solved correctly per pupil...... Per cent examples NOT solved correctly, per pupil	19 24	14 44	12 52	8 68
Middle group (122 pupils)	Av. No. examples solved correctly per pupil...... Per cent examples NOT solved correctly, per pupil	16 36	11 56	9 64	6 76
Lower group (54 pupils)	Av. No. examples solved correctly per pupil...... Per cent examples NOT solved correctly, per pupil	15 60	11 56	9 64	7 72
Entire grade (270 pupils)	Av. No. examples solved correctly per pupil...... Per cent examples NOT solved correctly, per pupil	17 32	12 52	10 40	7 72
GRADE 5					
Upper group (60 pupils)	Av. No. examples solved correctly per pupil..... Per cent examples NOT solved correctly, per pupil	18 28	14 44	13 48	10 40
Middle group (141 pupils)	Av. No. examples solved correctly per pupil...... Per cent examples NOT solved correctly, per pupil	18 28	21 16	12 52	10 40
Lower group (69 pupils)	Av. No. examples solved correctly per pupil..... Per cent examples NOT solved correctly, per pupil	17 32	11 32	11 56	9 64
Entire grade (270 pupils)	Av. No. examples solved correctly per pupil..... Per cent examples NOT solved correctly, per pupil	18 28	17 32	11 56	10 40

tory and the preliminary tests seem to parallel one another, yet the number of problems solved correctly in the inventory test is slightly greater than those solved correctly in the preliminary test, especially as the steps increase. This is to be expected if we assume that reasoning might tend to inhibit facility or accuracy in computation.

TABLE 4—(*Continued*)

Group		Problem Tests			
		One-Step	Two-Step	Three-Step	Four-Step
GRADE 6					
Upper group (55 pupils)	Av. No. examples solved correctly per pupil......	20	18	17	16
	Per cent examples NOT solved correctly, per pupil	20	28	32	36
Middle group (252 pupils)	Av. No. examples solved correctly per pupil.....	20	17	17	14
	Per cent examples NOT solved correctly, per pupil	20	32	32	44
Lower group (87 pupils)	Av. No. examples solved correctly per pupil......	19	16	15	13
	Per cent examples NOT solved correctly, per pupil	24	36	60	48
Entire grade (394 pupils)	Av. No. examples solved correctly per pupil......	20	17	16	14
	Per cent examples NOT solved correctly per pupil	20	32	36	44
GRADE 7					
Upper group (67 pupils)	Av. No. examples solved correctly per pupil.....	22	18	18	17
	Per cent examples NOT solved correctly, per pupil	12	28	28	32
Middle group (192 pupils)	Av. No. examples solved correctly per pupil......	21	18	18	16
	Per cent examples NOT solved correctly, per pupil	16	28	28	36
Lower group (57 pupils)	Av. No. examples solved correctly per pupil......	21	18	16	16
	Per cent examples NOT solved correctly, per pupil	16	28	36	36
Entire grade (316 pupils)	Av. No. examples solved correctly per pupil.....	22	18	17	16
	Per cent examples NOT solved correctly, per pupil	12	28	32	36

Table 4 is read as follows:—For Grade 3, there are 86 pupils in the upper group, having I.Q.s of 116 and upward; 137 pupils in the middle group with I.Q.s ranging from 100 to 115 inclusive; and 51 pupils in the lower group with I.Q. score of 99 or less. In the upper group the number of one-step problems which were solved correctly, out of the 25 problems given, averaged 15 per pupil; of the two-step problems which were entirely correct, we have an average of 13 problems per pupil; this means, according to the next line, that 40 per cent of the one-step problems and 48 per cent of the two-step problems per pupil were not solved correctly. Similar details are to be interpreted in like manner for each group in all grades, and for the combined groups of all grades tested.

Banting [1] in working directly with pupils in grades below the junior high school lists many causes of difficulty in problem solving. He states that one of the most fruitful sources of error is lack of ability to properly arrange written work in orderly, logical form, or care in thus arranging written work. Since this was already done for the child in the inventory tests, it would appear that the "fruitful" source of error was not greatly reduced.

Factoring Out Problem Errors Due Entirely to Computational Difficulties

In Table 5 the average number of problem errors per pupil which were made on the preliminary test are presented, by groups, for Grades 3, 4, 5, 6, and 7, after the problem errors due entirely to computational difficulties have been factored out. The procedure used in factoring out the problem errors for each child in the preliminary test that were due entirely to computational difficulties was that of subtracting the number of problems correct on the preliminary test from the total number of problems in the test, which was twenty-five. Then from this result subtract again the problem errors which were due entirely to computational difficulties. This leaves problem errors due to special causes other than pure computation. The total number of such problem errors for all one-, two-, three- or four-step problems are later correlated with reading ability in order to determine to what extent reading comprehension entered into the "remaining" errors made.

It will be seen that the errors not explained by computational difficulties entirely, increase from upper to lower groups. There is also a decided increase from one- to four-step problems. The percentages are noticeably rather high throughout.

It should be noted that for all groups in all grades there is a marked increase in errors, not explained by computational difficulties entirely, after the one-step problems. The same situation will be noted in the "I" and "C & I" columns of Table 6.

Index of Mastery and Classification of Problem Errors

Table 6 shows the Index of Mastery, by groups, and problem errors allocated to four different factors, for Grades 3, 4, 5, 6, and 7. The Index of Mastery is expressed in terms of per cent, and represents the amount or the degree to which unstable or erratic computation figures in. If a pupil performed his computations correctly in all three tests—preliminary, inventory, and final—no errors were charged against him in the way of erratic performance. However, if a pupil performed a certain computation correctly in both the preliminary and the inventory test, but failed on the same combination in the final test, an

TABLE 5

PROBLEM ERRORS IN THE PRELIMINARY TEST, BY GROUPS, FOR GRADES 3, 4, 5, 6, and 7, AFTER FACTORING OUT PROBLEM ERRORS DUE ENTIRELY TO COMPUTATIONAL DIFFICULTY

GROUP		PROBLEM TESTS				
		1-Step	2-Step	3-Step	4-Step	All Steps
GRADE 3						
Upper group (86 pupils)	Av. No. problem errors, per pupil NOT explained by computational difficulties entirely *.............	8	8			8
	Per cent problems due to errors not entirely computation **.................................	31	30			31
Middle group (137 pupils)	Av. No. prob. errors per pupil NOT ex. by comp. diff. entirely...................................	11	12			11
	Per cent problems due to errors not entirely comp...	45	46			46
Lower group (51 pupils)	Av. No. prob. errors per pupil NOT ex. by comp. diff. entirely...................................	12	11			11
	Per cent problems due to errors not entirely comp...	48	43			46
Entire grade (274 pupils)	Av. No. prob. errors per pupil NOT ex. by comp. diff. entirely...................................	10	10			10
	Per cent problems due to errors not entirely comp...	41	41			41
GRADE 4						
Upper group (94 pupils)	Av. No. prob. errors per pupil NOT ex. by comp. diff. entirely...................................	5	13	13	15	12
	Per cent problems due to errors not entirely comp...	19	52	53	61	47
Middle group (122 pupils)	Av. No. prob. errors per pupil NOT ex. by comp. diff. entirely...................................	7	15	16	18	12
	Per cent problems due to errors not entirely comp...	24	60	65	73	48
Lower group (54 pupils)	Av. No. prob. errors per pupil NOT ex. by comp. diff. entirely...................................	8	16	17	20	15
	Per cent problems due to errors not entirely comp...	33	63	68	78	60
Entire grade (270 pupils)	Av. No. prob. errors per pupil NOT ex. by comp. diff. entirely...................................	6	14	15	17	13
	Per cent problems due to errors not entirely comp...	26	58	62	70	50
GRADE 5						
Upper group (60 pupils)	Av. No. prob. errors per pupil NOT ex. by comp. diff. entirely...................................	5	9	11	13	9
	Per cent problems due to errors not entirely comp...	19	37	43	52	38
Middle group (141 pupils)	Av. No. prob. errors per pupil NOT ex. by comp. diff. entirely...................................	7	13	16	17	13
	Per cent problems due to errors not entirely comp...	28	51	63	69	53
Lower group (69 pupils)	Av. No. prob. errors per pupil NOT ex. by comp. diff. entirely...................................	7	14	16	19	14
	Per cent problems due to errors not entirely comp...	28	55	64	75	55
Entire grade (270 pupils)	Av. No. prob. errors per pupil NOT ex. by comp. diff. entirely...................................	6	12	15	17	13
	Per cent problems due to errors not entirely comp...	26	49	59	67	50

TABLE 5—(*Continued*)

Group		Problem Tests 1-Step	2-Step	3-Step	4-Step	All Steps
GRADE 6						
Upper group (55 pupils)	Av. No. prob. errors per pupil NOT ex. by comp. diff. entirely	3	7	9	12	8
	Per cent problems due to errors not entirely comp.	11	28	38	47	31
Middle group (252 pupils)	Av. No. prob. errors per pupil NOT ex. by comp. diff. entirely	5	10	13	15	11
	Per cent problems due to errors not entirely comp.	19	41	53	61	43
Lower group (87 pupils)	Av. No. prob. errors per pupil NOT ex. by comp. diff. entirely	7	13	16	16	13
	Per cent problems due to errors not entirely comp.	27	50	63	64	52
Entire grade (394 pupils)	Av. No. prob. errors per pupil NOT ex. by comp. diff. entirely	5	10	13	15	15
	Per cent problems due to errors not entirely comp.	20	41	53	60	59
GRADE 7						
Upper group (67 pupils)	Av. No. prob. errors per pupil NOT ex. by comp. diff. entirely	2	5	6	10	6
	Per cent problems due to errors not entirely comp.	7	19	25	40	23
Middle group (192 pupils)	Av. No. prob. errors per pupil NOT ex. by comp. diff. entirely	3	8	10	13	8
	Per cent problems due to errors not entirely comp.	14	32	38	50	33
Lower group (57 pupils)	Av. No. prob. errors per pupil NOT ex. by comp. diff. entirely	5	10	13	15	11
	Per cent problems due to errors not entirely comp.	21	41	51	58	43
Entire grade (316 pupils)	Av. No. prob. errors per pupil NOT ex. by comp. diff. entirely	3	8	9	12	8
	Per cent problems due to errors not entirely comp.	14	31	35	49	32

* This was determined by averaging all problem errors, which were found by subtracting the number of problems correct on the preliminary test from the total number of problems which each pupil worked or attempted (25). Then from this result subtract again the errors which were due entirely to computational difficulties.

** Total number of problem errors for the group, not explained by computational difficulties entirely ÷ (25 × number of pupils).

Table 5 is read as follows:—For Grade 3, there are 86 pupils in the upper group, having I.Q.s of 116 and upward; 137 pupils in the middle group with I.Q.s ranging from 100 to 115 inclusive; and 51 pupils in the lower group with I.Q. scores of 99 and less. In the upper group the average number of one-step problem errors in the preliminary test which are due to causes other than computation entirely was 8 per pupil; of the two-step problems also 8 per pupil; for the combined or total problem errors in both one-step and two-step problems 8 per pupil. In other words (according to the remaining figures in the upper group), 31 per cent of the one-step problem errors in the preliminary test are due to causes other than computation entirely; 30 per cent of the two-step problems; and 31 per cent of the combined problem errors of both the one-step and two-step problems. Similar details are to be interpreted in like manner for each group in all grades, and for the total groups of all grades tested.

TABLE 6

INDEX OF MASTERY, BY GROUPS, AND PROBLEM ERRORS ALLOCATED TO FOUR DIFFERENT FACTORS, FOR GRADES 3, 4, 5, 6, AND 7

GROUP		PROBLEM TESTS															
		ONE-STEP				TWO-STEP				THREE-STEP				FOUR-STEP			
	*	C.	I.	C. & I.	E.	C.	I.	C. & I.	E.	C.	I.	C. & I.	E.	C.	I.	C. & I.	E.
GRADE 3																	
Upper group	Av. No. problem errors per pupil	4	2	2	9	4	2	3	10								
(86 pupils)	Index Mastery **				35				39								
Middle group	Av. No. problem errors per pupil	4	2	5	9	4	3	6	9								
(137 pupils)	Index Mastery				34				35								
Lower group	Av. No. problem errors per pupil	4	3	4	9	4	3	6	8								
(51 pupils)	Index Mastery				35				36								
Entire grade	Av. No. problem errors per pupil	4	2	4	9	4	2	5	9								
(274 pupils)	Index Mastery				35				36								
GRADE 4																	
Upper group	Av. No. problem errors per pupil	2	.7	.5	8	3	4	5	8	3	4	7	8	4	3	10	6
(94 pupils)	Index Mastery				33				33				32				24
Middle group	Av. No. problem errors per pupil	3	1	2	9	3	4	8	7	6	4	10	6	3	3	13	5
(122 pupils)	Index Mastery				38				29				23				19
Lower group	Av. No. problem errors per pupil	2	2	3	10	3	4	8	8	2	3	11	6	2	3	14	5
(54 pupils)	Index Mastery				40				32				23				19
Entire grade	Av. No. problem errors per pupil	2	1	2	9	3	4	7	8	5	4	9	6	3	3	11	5
(270 pupils)	Index Mastery				36				31				26				21
GRADE 5																	
Upper group	Av. No. problem errors per pupil	2	1	1	9	5	3	2	9	5	3	4	10	5	4	7	8
(60 pupils)	Index Mastery				36				38				41				30
Middle group	Av. No. problem errors per pupil	2	2	1	9	3	4	5	8	3	5	7	7	3	5	10	5
(141 pupils)	Index Mastery				37				33				29				21
Lower group	Av. No. problem errors per pupil	2	2	2	10	3	5	6	8	3	5	8	7	3	5	11	5
(69 pupils)	Index Mastery				37				34				32				23
Entire grade	Av. No. problem errors per pupil	2	2	1	9	3	4	5	8	4	4	7	8	4	5	9	6
(270 pupils)	Index Mastery				37				34				32				23

TABLE 6 (*Continued*)

Group		Problem Tests															
		One-Step				Two-Step				Three-Step				Four-Step			
		C.	I.	C. & I.	E.	C.	I.	C. & I.	E.	C.	I.	C. & I.	E.	C.	I.	C. & I.	E.
GRADE 6																	
Upper group (55 pupils)	Av. No. problem errors per pupil	1	1	.4	5	2	2	1	9	2	3	1	10	3	4	4	9
	Index Mastery................				21				38				41				34
Middle group (252 pupils)	Av. No. problem errors per pupil	1	1	.4	7	3	3	2	9	3	5	3	9	4	5	6	9
	Index Mastery................				30				35				36				35
Lower group (87 pupils)	Av. No. problem errors per pupil	1	2	1	8	3	5	4	8	2	6	6	7	3	6	8	6
	Index Mastery................				33				33				29				24
Entire grade (394 pupils)	Av. No. problem errors per pupil	1	1	1	7	3	4	3	9	3	5	3	9	3	5	7	8
	Index Mastery................				29				35				35				33
GRADE 7																	
Upper group (67 pupils)	Av. No. problem errors per pupil	.4	1	.1	5	1	1	.4	7	1	2	1	8	1	4	2	8
	Index Mastery................				18				29				32				34
Middle group (192 pupils)	Av. No. problem errors per pupil	1	1	.2	6	1	3	1	9	2	3	2	8	2	5	3	8
	Index Mastery................				25				34				34				33
Lower group (57 pupils)	Av. No. problem errors per pupil	1	1	1	7	2	4	2	9	2	4	4	9	2	6	5	8
	Index Mastery................				28				36				34				31
Entire grade (316 pupils)	Av. No. problem errors per pupil	1	1	.2	6	1	3	1	8	2	3	2	8	2	5	3	8
	Index Mastery................				24				34				33				33

*"C" = Errors due entirely to computational difficulties.
"I" = Errors due entirely to inherent difficulties.
"C&I" = Errors due entirely to computational and inherent difficulties.
"E" = Errors due entirely to erratic performance.
** Per cent of the total number of errors to the number of attempts at problems (25 % number of pupils).

Table 6 is read as follows:—For Grade 3, there are 86 pupils in the upper group, having I.Q.s of 116 and upward; 137 pupils in the middle group with I.Q.s ranging from 100 to 115, inclusive; and 51 pupils in the lower group with I.Q. scores of 99 and less. In the upper group the average number of problem errors per pupil in the one-step problems, due entirely to computational difficulties, is 4; of the two-step problems the average is also 4 per pupil. The average number of problem errors per pupil in the one-step problems, due entirely to inherent difficulties, is 2; of the two-step problems the average is also 2 per pupil. The average number of problem errors per pupil in the one-step problems which may be attributed to both computational and inherent difficulties is 2; of the two-step problems the average is 3. The average number of problem errors per pupil in the one-step problems which may be attributed to erratic performance entirely is 9; of the two-step problems the average is 10. The Index of Mastery in the one-step problems for the upper group is 35, or in other words, 35 is the per cent of the total number of erratic problem errors for the upper group to the number of attempts which were made at the problems (twenty-five times the number of pupils). For the two-step problems the Index of Mastery is 39. Similar details are to be interpreted in like manner for each group in all grades, and for the combined groups of all grades tested.

erratic error was charged against him. The same principle held true with any combination of the three tests given (see Chart I). The Index of Mastery was determined by totaling all of the erratic performance errors for each group or grade and then dividing by the total number of attempts which were made at the problems, which would be twenty-five times the number of pupils. The four different factors to which the problem errors were allocated were (1) computational difficulties entirely, (2) inherent difficulties, (3) computational and inherent difficulties, and (4) erratic performance only. An examination of Chart I will serve to explain how these factors were arrived at.

It should be noted that the "I" and "C & I" columns may include problems that could not be worked by the pupils and also problems that were incomplete, because all such problems when recorded on the record sheets were accounted as "*wrong.*"

The errors set forth in columns "I" and "C & I" consistently increase from the upper to lower groups and also from one-step problems to four-step problems. The increase from upper to lower groups may show that intelligence consistently operates.

The figures in column "I" show a decided increase in errors due entirely to inherent difficulties for all groups, after the one-step problems. This may indicate that beyond one-step problem work pupils may lack reasoning techniques in applying their computational ability to the solution of verbal problems. In the one-step problems, the number of errors due entirely to inherent difficulties are rather small and seemingly of equal importance for all grades. This may point to the fact that the solution of one-step problems is not so much a matter of reasoning as it is an automatic response to similar problems that have been solved, or to straight computational ability.

In examining the Index of Mastery in column "E," for one-step problems, there is a decided and consistent percentage increase from upper to lower groups for all grades. This is to be expected if we assume that pupils with a lower I.Q. do make a larger number of mistakes than those with a higher I.Q. However, it should be noted that after the one-step problems there is a decided decrease in the percentage of mastery from upper to lower groups, especially as the steps increase. Just the opposite is naturally expected. The answer lies in several possibilities. First, the brighter pupils are probably confronted with a greater number of choices in solving their problems, hence the possibility of a greater erratic performance. Second, the brighter pupils probably did not give up as quickly as the slower pupils did but went straight through and tried all the problems with greater varying success and with better computational ability, thus making for greater erratic performance. Third, there were possibly greater disturbances on the part of the slower pupils than the brighter pupils when facing and solving the problems. In handwriting we know that the quality becomes poorer as one attends to the thought being expressed. The same may be true in problem solving. Hence there may be a greater erratic performance on the part of the slower pupils.

Statistical Correlations

In order to get a clearer picture of the relationships between certain factors involved in this study and in order to shed further light on the data presented in the foregoing tables, a series of statistical correlations were made.

Paragraph Comprehension and Problem Errors

Table 7 shows the correlations between paragraph comprehension scores and problem errors which were made on the preliminary tests. The problem errors were those not explained by computational difficulties entirely.

TABLE 7

Correlations Between Paragraph Comprehension Scores and Problem Errors Made on the Preliminary Tests, Not Explained by Computational Difficulties Entirely

Grade	Upper Group	Middle Group	Lower Group	Total Group
Grade 3	−.4786	−.2388	−.0537	−.4170
Grade 4	−.4107	−.2899	−.0058	−.3535
Grade 5	−.0879	−.2366	−.1272	−.2107
Grade 6	−.1331	−.1922	−.0135	−.2992
Grade 7	−.1589	−.3011	−.3712	−.4761

All the correlations in this table are expected to be negative, since the problem error scores for each child were found by subtracting the number of problems correct on the preliminary test from the total number of problems in the test (which was

TABLE 8

CORRELATIONS RELATING TO GENERAL INTELLIGENCE AND GAINS MADE BETWEEN THE PRELIMINARY AND FINAL PROBLEM TESTS

GRADE	UPPER GROUP		MIDDLE GROUP		LOWER GROUP		TOTAL GRADE	
	M. A. & Gains	I. Q. & Gains	M. A. & Gains	I. Q. & Gains	M. A. & Gains	I. Q. & Gains	M. A. & Gains	I. Q. & Gains
One-Step Problems								
3	.0281	−.0318	−.1556	−.0085	.1472	.2567	−.0488	−.0049
4	−.1668	.1580	.2084	.0564	−.0026	.0141	.0202	.0046
5	−.2148	−.2166	−.1014	−.1346	−.0917	.0557	−.2285	−.2136
6	−.1540	.0224	−.0825	−.0397	−.1950	.0116	−.1470	−.0906
7	−.2937	−.2753	−.2015	−.1747	−.1727	−.1403	−.3023	−.2859
Two-Step Problems								
3	.0678	−.0118	−.0504	.0078	.2930	.1070	.0752	.0494
4	.1267	−.0603	.1146	.0616	.1462	−.0657	.1915	.1397
5	−.1032	−.0166	−.1405	−.0836	.0038	−.0941	−.0597	−.0121
6	−.3972	−.2114	−.1212	−.0984	−.1417	−.0515	−.0642	−.0276
7	−.1758	.0290	−.0220	−.0952	−.0229	−.0786	−.0989	−.1152
Three-Step Problems								
3	....	....	....	....	....	....	....	....
4	.2130	.0551	.1571	.1289	.0805	−.0369	.1710	.0893
5	.0275	.0967	−.1935	−.1268	.0038	−.2213	−.0294	.0018
6	−.3921	−.1809	−.0500	−.0830	−.0309	−.0801	−.0334	−.0114
7	−.0290	.0941	.0048	.0550	.0711	.0411	−.0523	−.0382
Four-Step Problems								
3	....	....	....	....	....	....	....	....
4	.1660	.0255	.1152	.1172	.0613	.0511	.1732	.1286
5	−.1421	.0212	−.0485	−.0819	−.2014	−.2660	−.0575	.0078
6	−.2164	−.0835	−.0770	−.0489	−.0475	−.0068	−.0218	.0110
7	−.0770	−.0849	−.0509	.2038	.0395	−.1747	.0001	−.0669

twenty-five); then from this result again subtracting the problem errors which were due entirely to computational difficulties.

For Grades 3 and 4 the correlations decrease from upper to lower groups. In Grade 7 the opposite situation occurs; the correlations increase from upper to lower groups.

General Intelligence and Gains Made

Table 8 shows the correlations between the mental ages and the gains made between the preliminary and the final problem tests; and also between the intelligence quotients and the gains made between the preliminary and the final problem tests. All the correlations in this table are low, with a preponderance of negative rather than positive correlations. The negative correlations result in larger numbers than those of the positive correlations. There are a smaller number of negative correlations in the third and fourth grades than in the fifth, sixth, or seventh grades. Comparing the negative correlations of the upper groups with those of the lower groups, it is evident that the negative correlations of the upper groups are much larger than those of the lower groups. This is true for all grades, which would indicate that the lower groups made a more consistent computational improvement than did the upper groups.

M.A., I.Q., Inventory Test, and Preliminary Test

Table 9 shows a set of three correlations for each of the upper, middle, and lower groups in Grades 3 to 7 inclusive, relating to one-, two-, three-, and four-step problems. The set of three correlations relates to the mental age (M.A.), intelligence quotient (I.Q.), and the inventory test as correlated with the preliminary test.

In following through the correlations made between the inventory test and the preliminary test, it will be observed that in the one-step problems the correlations become increasingly higher for all grades as one goes from the upper to the lower groups. The only exception to this appears in the lower group of Grades 4 and 7. The correlations in the sixth grade become increasingly higher for all of the one-, two-, three-, and four-step problems as one goes from the upper to the lower groups. The same is true for Grade 3. In Grade 5 the correlations increase for all of one-, two-, three-, and four-step problems as one goes

TABLE 9

Correlations

Grade	Upper Group			Middle Group			Lower Group			Total Grade		
	Inv.* & P. T.	M. A.* & P. T.	I. Q.* & P. T.	Inv. & P. T.	M. A. & P. T.	I. Q. & P. T.	Inv. & P. T.	M. A. & P. T.	I. Q. & P. T.	Inv. & P. T.	M. A. & P. T.	I. Q. & P. T.
Correlations Relating to *One-Step Problems*												
3	.2228	.2588	.3023	.3197	.3585	.2492	.5470	−.0187	.2348	.3531	.3677	.3742
4	.3985	.3389	−.0491	.4757	−.0638	.1948	.1965	.2184	.0655	.4275	.3122	.2824
5	.4416	.2828	.1907	.5309	.3243	.2506	.5482	.1618	.0825	.5155	.3253	.2494
6	.1494	.1588	.0348	.2070	.2468	.1861	.4404	.2183	−.0256	.2419	.3872	.3320
7	.3232	.3812	.2407	.3736	.3560	.2605	.2267	.3954	.3841	.3235	.4901	.4543
Correlations Relating to *Two-Step Problems*												
3	.1927	.2246	.0780	.1670	.1494	.1700	.4471	−.1422	.0774	.2754	.2369	.2907
4	.4533	.2044	.2337	.3786	.0628	.1517	.3208	.2020	.0163	.4126	.2213	.1814
5	.3611	.3261	.2184	.4749	.3177	.1327	.3172	.0783	.0327	.4219	.3013	.1986
6	.2089	.4426	.2697	.3337	.2652	.2065	.5718	.1847	.0242	.3966	.4010	.3404
7	.3192	.3711	.2043	.3495	.2654	.2355	.3885	.3615	.2792	.3241	.4684	.4360

*Inv.—Inventory Test
P. T.—Preliminary Test

M. A.—Mental Age
P. T.—Preliminary Test

I. Q.—Intelligence Score
P. T.—Preliminary Test

TABLE 9—(*Continued*)

Grade	Upper Group			Middle Group			Lower Group			Total Grade		
	Inv.* & P. T.	M. A.* & P. T.	I. Q.* & P. T.	Inv. & P. T.	M. A. & P. T.	I. Q. & P. T.	Inv. & P. T.	M. A. & P. T.	I. Q. & P. T.	Inv. & P. T.	M. A. & P. T.	I. Q. & P. T.
Correlations Relating to *Three-Step Problems*												
3	....	....	....	....	....	....	....	....	....	....	....	....
4	.4206	.1626	.1303	.3729	.0799	.1268	.5730	.2466	−.0286	.4665	.2824	.2269
5	.3415	.2685	.1740	.4269	.3331	.1330	.1012	.0742	−.0140	.3407	.3062	.2020
6	−.0391	.3628	.0458	.1146	.2855	.2205	.3553	.2426	.0801	.1622	.4162	.3413
7	.4321	.3996	.2451	.3288	.2649	.2153	.2576	.2657	.1774	.3276	.4622	.4255
Correlations Relating to *Four-Step Problems*												
3	....	....	....	....	....	....	....	....	....	....	....	....
4	.4772	.2555	.1489	.2755	.0714	.0592	.5706	.3070	.0756	.4220	.2953	.2480
5	.2854	.2776	.1753	.3005	.3028	.1280	.0498	.1345	−.0413	.2364	.3353	.2325
6	.0486	.1963	.0627	.1465	.2407	.2047	.4641	.1947	.0260	.2408	.3763	.3243
7	.3591	.5115	.3321	.2917	.1872	.1845	.3859	.2997	.1938	.3213	.4572	.4299

*Inv.—Inventory Test
P. T.—Preliminary Test

M. A.—Mental Age
P. T.—Preliminary Test

I. Q.—Intelligence Score
P. T.—Preliminary Test

from the upper to the middle groups only. In Grade 7 the correlations increase from the upper to the middle group on one- and two-step problems, but decrease on three- and four-step problems. When correlating the scores for the entire grades, we note a consistent increase from Grade 3 to 5 inclusive. Beyond this point where the fifth grade tests were given to the sixth and seventh grades, we note a consistent falling off for the sixth and seventh grades, with the exception of the correlation made by the seventh grade on the four-step problems.

Comparing the M.A. and I.Q. correlations with the preliminary tests, it will be observed that the correlations between the M.A. scores and the preliminary test scores are decidedly higher than those of the I.Q. scores and the preliminary test scores, in nearly all grades and all groups. This would indicate that M.A. is a better basis for problem classification. The only exceptions to this occur in the third and fourth grades. In Grade 3 the I.Q. correlations are higher than the M.A. correlations for the upper and lower groups in one-step problems; and for the middle and lower groups in the two-step problems. In Grade 4 the I.Q. correlations are higher than the M.A. correlations for the middle group in the one-step problems; for the upper and middle groups in the two-step problems; and for the middle group in the three-step problems. An inspection of M.A. and I.Q. correlations with the preliminary test scores shows that there are no consistent correlations between groups or within groups in going from one-step problems to four-step problems. But when noting the correlations for the entire grades, it should be observed that, with the exception of Grade 3, the M.A. correlations increase from Grade 4 to 7 inclusive, on all problem steps. The same tendency is true for the I.Q. correlations.

From an examination of all the correlations computed, it will be seen that the inventory and preliminary test correlations range the highest. The only exceptions to this appear in the correlation established between paragraph comprehension and problem errors in the preliminary tests not due to computational difficulties entirely. In the latter instance, the lower group of Grades 5 and 7 ranks higher in three- and four-step problems and in one- and three-step problems respectively; the middle group of Grades 3, 4, 6, and 7 ranks higher in two-step, four-step, three- and four-step, and four-step problems respectively; the upper group of Grades 3, 4, and 6 ranks higher in one- and two-step, one-step, and three- and four-step problems respectively; and the entire Grades 3 and 7, including the one-step, three-step, and four-step problems of Grades 6 rank somewhat higher. This substantiates the fact that computational difficulties figure into "problem solving" rather formidably.

Problem Performance by I.Q. Groups

The following graphs show the overlapping of arithmetic scores made by three non-overlapping I.Q. groups in each grade on three-step problems which were solved correctly in the final tests. The graphs show that in spite of the non-overlapping of the I.Q. groups, each I.Q. group in each grade has the same range in arithmetic scores performance. That is, there were some

GRAPH A

GRADE 4: Showing Overlapping of Arithmetic Scores on Three-Step Final Test Problems for Three Non-Overlapping I.Q. Groups

Per cent of pupils receiving each score

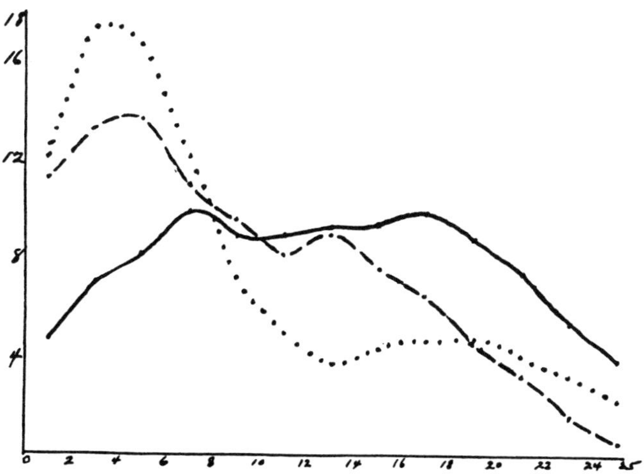

Number of problems solved correctly

	Range of I.Q.	No. of Pupils	Key
Group 1	116 and up	94	———
" 2	100-115 inclusive	121	—.—.—.
" 3	99 and less	54	......

pupils in each of the I.Q. groups for each grade who solved none of the three-step problems correctly, and some pupils in each of the I.Q. groups for each grade who solved all of the twenty-five problems correctly. Thus it would appear that homogeneous

GRAPH B

GRADE 5: Showing Overlapping of Arithmetic Scores on Three-Step Final Test Problems for Three Non-Overlapping I.Q. Groups

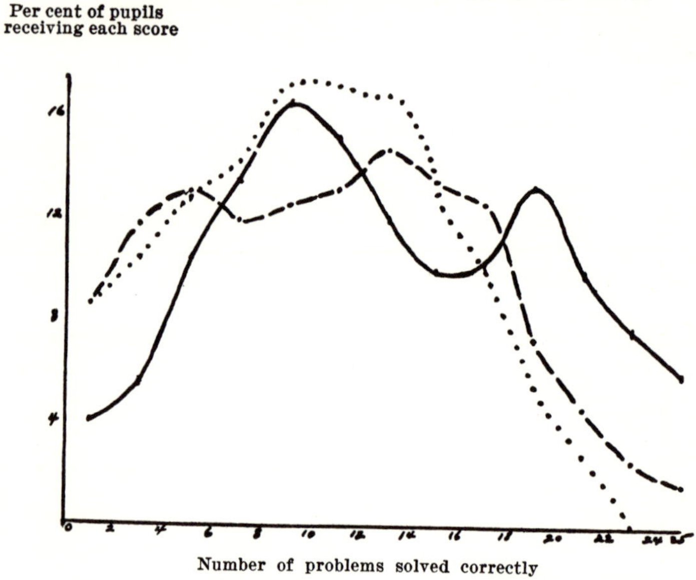

	Range of I.Q.	No. of Pupils	Key
Group 1	116 and up	60	———
" 2	110-115 inclusive	140	–.–.–.
" 3	99 and less	66	......

grouping on the basis of I.Q. does not produce homogeneity for problem solving. However, it should be noted that in almost every case, the per cent of pupils in the lower I.Q. groups solving none of the problems correctly is greater than the per cent of those for the middle and upper I.Q. groups, and that the per cent of pupils in the upper I.Q. groups solving all of the problems correctly is greater than the per cent of those for the middle and upper groups.

GRAPH C

GRADE 6: Showing Overlapping of Arithmetic Scores on Three-Step Final Test Problems for Three Non-Overlapping I.Q. Groups

Per cent of pupils receiving each score

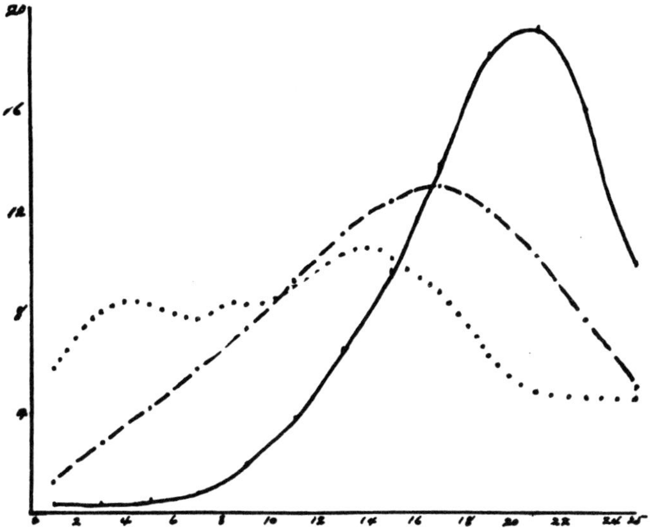

Number of problems solved correctly

	Range of I.Q.	No. of Pupils	Key
Group 1	116 and up	55	———
" 2	110-115 inclusive	251	–.–.–.
" 3	99 and less	87	......

GRAPH D

Grade 7: Showing Overlapping of Arithmetic Scores on Three-Step Final Test Problems for Three Non-Overlapping I.Q. Groups

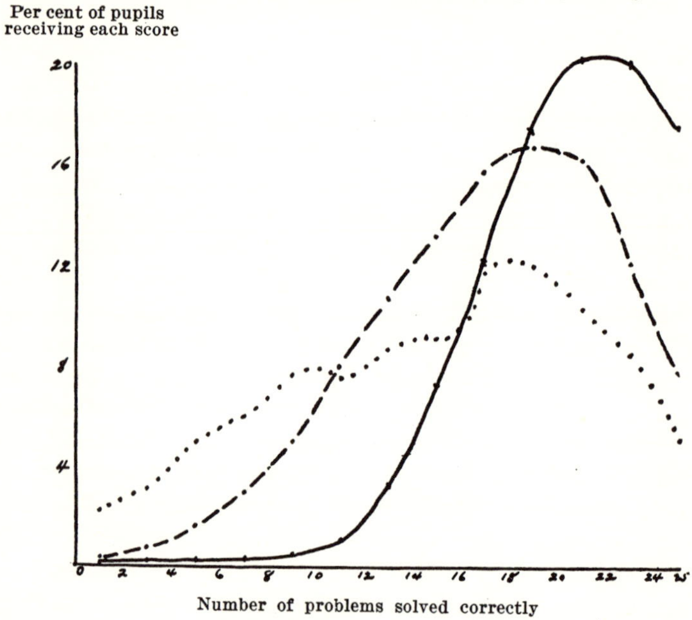

	Range of I.Q.	No. of Pupils	Key
Group 1	116 and up	65	–.–.–.
" 2	110-115 inclusive	187	– – –
" 3	99 and less	57	……

CHAPTER VI

CONCLUSIONS

General Statement

In this study no attempt has been made to generalize except within the limits justified by the evidence presented. No attempt has been made to deal with coefficients of correlations because in doing so numerous difficulties would be encountered. For example, in a study of certain factors involved in problem solving we know at the outset that a positive relationship exists between the more important paired facts such as the relationship between intelligence expressed in M.A. or I.Q. scores, and problem-solving ability. What we might wish to know or what we do not know is how nearly all this relationship approaches perfection. Any coefficient of correlation expressing the degree of perfection of such an agreement would not lead to satisfactory interpretations, because it is impossible for us to specify the population differing from the population tested only by the chance of selection. Accordingly, it would mean very little to show, for example, that a given coefficient of correlation fifty or one hundred times its probable error is due to sampling.[1]

Again, when we attempt to study, for example, the relationship between reading comprehension and errors due entirely to computational difficulties we encounter the difficulty of interpreting probable errors because no appropriate allowances have been made for the non-representativeness of the data. Accordingly, probable errors are not set forth for interpretative purposes in this study. If it were possible to define exactly or approximately the population tested, the formula for probable error of sampling immediately becomes superfluous.

Furthermore, we know that the arithmetical performance of grade school pupils is very erratic. Low correlations, therefore,

[1] Monroe, W. S., "Difficulties in Educational Research," *Educational Research Bulletin, Ohio University,* January 8, 1930, pp. 1-6.

between any two factors, such as scores on the inventory tests and scores on the preliminary tests, could not be considered as being especially significant.

General Conclusions

No study would be complete which evaded the purposes set forth at the outset. The major concern of the writer has been to determine as accurately as possible within the limitations of this study the relative influence of each purpose as defined at the outset. They may be summarized as follows:

1. There is wide diversity of practice in the number of steps in problem materials recommended for use in the twenty-four outstanding courses of study examined. The allocation of one-step problems in these courses of study range from Grade 1 to 4; two-step problems from Grade 2 to 5. One-step problems are largely allocated to the second or third grade; two-step problems to the fourth grade; three-step problems to the fifth or sixth grade; and problems of more than three steps to the sixth grade.

2. If we set up a standard of only 60 per cent mastery or achievement in problem solving for an entire grade after drill has been given on computational difficulties, we find that the beginning third, fourth, and fifth grade pupils achieve this on only two-step problems. In the sixth and seventh grades, where the fifth grade problems were administered, we find approximately a 60 per cent achievement or better on one-, two-, three-, and four-step problems for all groups.[2] Had the computational difficulties increased with each of Grades 6 and 7, it is probable that very little more than one- or two-step problems would have been mastered. The lower group in the sixth grade apparently did not do as well as the lower group in the fifth grade. The upper group in the seventh grade was able to solve four-step problems of fifth grade difficulty with 68 per cent mastery or achievement.

3. The correlation between reading ability and arithmetic errors in problem solving not explained by computational difficulties ranges from .2107 for the fifth grade to .4761 for the seventh grade. The average correlation for all grades, Grades 3 to 7 inclusive, is .3513. In Grades 3 and 4 the correlations

[2] See Appendix, page 94a, for modification of these statements.

decrease from upper to lower groups. In Grade 7 the correlations increase from upper to lower groups.

4. With the exception of the three- and four-step problems in Grade 4, the four-step problems in Grade 5, and the one-step problems in Grades 6 and 7, the Index of Mastery ranges from 31 to 37 per cent for all grades on all problem steps studied. The average Index of Mastery for all grades on all problem steps, excluding the exceptions mentioned above, is 34.1 per cent.

5. The data reveal no consistent correlations between problem-solving ability and intelligence levels as determined by upper, lower, and middle groups. The I.Q. correlations for Grades 3 to 7 inclusive range from —.0491 to .3321 for the upper group; from —.0592 to .2506 for the middle group; and from —.0413 to .3841 for the lower group.

Using intelligence levels to mean entire grades, we find that with the exception of Grade 3, all M.A. and P.T. correlations increase from Grades 4 to 7 inclusive, on all problem steps. The same is true for the I.Q. and P.T. correlations, with the exception of Grade 5. The greatest range, by grades for both M.A. and I.Q. correlations appear in the two-step problems. Here we find the M.A. correlations ranging from .2213 in the fourth grade to .4684 in the seventh grade; the I.Q. correlations ranging from .1814 in the fourth grade to .4360 in the seventh grade.

Interpreting M.A. and I.Q. correlations as representing different intelligence levels, we find that the correlations between M.A. scores and the preliminary test scores are decidedly higher than those of the I.Q. scores and the preliminary test scores in nearly all grades and all groups.

6. There is a decidedly closer relationship between problem-solving ability and skill in computation than between problem-solving ability and general intelligence for all groups in Grades 4 and 5 inclusive. The same is true for all of the lower groups in Grades 3 and 6, and for the middle group in Grade 7. Problem performance by entire grades shows that in Grade 3 the relationship is about equal; in Grades 4 and 5 decidedly greater; and in Grades 6 and 7, where problems of fifth grade difficulty were given, considerably less.

7. "Dependable habits of computation" seem essential to success in problem solving. This is brought out by a study of

Tables 4 and 6, which set forth "erratic performance" and "inventory test scores."

8. Improvement in computational accuracy or drill in mastering the essential computations did increase the ability to earn scores on verbal problem tests. Results are set forth in Table 3.

9. A preponderance of negative correlations exists between intelligence scores and gains made between the preliminary tests and the final tests in problem solving. The negative correlations result in larger numbers than those of the positive correlations. The negative correlations of the upper groups are much larger than those of the lower groups.

Other general conclusions which have developed in this study may be stated as follows:

a. After one- or two-step problems, it appears that the number of steps do not greatly influence the results of problem solving.

b. That according to our present method of measuring problem-solving ability, intelligence or reading ability does not formidably enter in.

c. The tendency in certain schools, especially in those where considerable attention is given to projects and activities, to decrease the amount of time given to computation, seems destined to lead to disappointing results.

d. The drill upon computation in Grades 5, 6, and 7 must not be relaxed if satisfactory conditions in problem solving are to be achieved.

e. Accuracy in computation is basic to the solution of verbal problems; the ability to earn scores on verbal problem tests increases as soon as good computation habits are established.

f. As the steps in problem solving increase, the failure to work problems correctly may possibly be attributed as largely to unstable computational ability as to a "lack of reasoning ability" or a lack of reasoning techniques. This is borne out by the results of the inventory test and also from the fact that the per cent of problems NOT solved correctly in the inventory tests seem to parallel or approximate the per cent of problems NOT solved correctly in the preliminary tests.

g. Reading comprehension is not as significant, statistically, as the factor of computation in problem-solving results.

Although the data of this investigation are not entirely consistent, they appear to substantiate the conclusion that one of the universal weaknesses or difficulties in our problem-solving work is the fact that we have attempted to measure problem-solving ability in terms of getting the right answer in place of correct reasoning abilities. Fallacies may result from either emphasis. If we say that one school solves problems much better than another school, we may merely mean that one school is more skillful in computation than another school, since "answer getting" does depend upon effective computation. This fact has already been brought out in this study. On the other hand, it is entirely conceivable that a pupil may be excellent in his ability to restate a problem in his own words, to note relationships, to select the right processes, or to make the necessary "short cuts," all of which are indicative of genuine reasoning, yet fail to "get the right answer" because of computational difficulties or erratic performance.

Our data reveal the fact that in the preliminary tests, third-grade pupils solve two-step problems with greater facility than one-step problems. Yet in the succeeding grades we have the following situation. Two-step problems are worked only 50 per cent as correctly in the fourth grade; 56 per cent as correctly in the fifth grade; 63 and 76 per cent as correctly in the sixth and seventh grades respectively, where verbal problems of only fifth grade difficulty were given. This would indicate that our schools have accomplished very little in the successful teaching of problem solving; that a large per cent of grade school pupils suddenly fail to make the effort to do reflective thinking after leaving the third grade, or that many pupils have simply failed in computing the right answer or in making the right solution. Such results also indicate that our present practice relative to problem solving may be based on the failure of teachers to understand what reasoning is or the factors which go to make up reasoning in problem solving. It is very probable that neither pupils nor teachers know what reasoning is, and accordingly are unable to achieve desirable results. Like the teaching of history, we are possibly insisting upon statements of facts or "answers" rather than upon ability to see relationships or to formulate ideas based upon materials that have already been learned.

BIBLIOGRAPHY

1. BANTING, G. O. "The Elimination of Difficulties in Reasoning." *Second Yearbook of the Department of Elementary School Principals.* National Education Association, Washington, D. C., 1923, pp. 411-21.
2. BONSER, FREDERICK G. D. *The Reasoning Ability of Children of the Fourth, Fifth, and Sixth School Grades.* Contributions to Education, No. 37. Bureau of Publications, Teachers College, Columbia University, 1910.
3. BRADFORD, E. J. G. "Suggestion, Reasoning and Arithmetic," *Forum of Education,* 3:3-12, February, 1925.
4. BUCKINGHAM, B. R. "Selecting and Evaluating Arithmetic Problems," *Educational Research Bulletin,* No. 4. (Ohio State University), 1925, pp. 34-37.
5. BUSWELL, GUY T. and JUDD, C. H. *Summary of Educational Investigations Relating to Arithmetic.* Supplementary Monograph No. 27. University of Chicago, 1925.
6. CLARK, JOHN R. and VINCENT, LEONA. "A Comparison of Two Methods of Analyzing Problems." *Mathematics Teacher,* 8:226-33, April, 1925.
7. DURELL, FLETCHER. "Solving Problems in Arithmetic." *School Science and Mathematics.* 28:925-35, December, 1928.
8. FISHER, S. CAROLYN. "Arithmetic Reasoning in Children," *Pedagogical Seminary,* 19:48-77, March, 1912.
9. GRAHAM, VERNE O. "Arithmetic Reasoning Project and the Measurement of Improvement." *Second Yearbook.* Chicago Principals' Club, 1927.
10. GREENE, HARRY A. "Directed Drill in the Comprehension of Verbal Problems in Arithmetic." *Journal of Educational Research,* 11:33-40, January, 1925.
11. HAMILTON, E. R. "Insight and Skill in Arithmetic." *Journal of Educational Research,* 12:136-44, September, 1925.
12. HANNA, PAUL R. *Arithmetic Problem Solving.* A Lincoln School Publication. Bureau of Publications, Teachers College, Columbia University, 1929.
13. HECK, W. H. "The Efficiency of Grammar Grade Pupils in Reasoning Tests in Arithmetic at Different Periods of the School Day." *Journal of Educational Psychology,* 5:92-95, February, 1914.
14. HERR, ROSS. "Rationalization in Arithmetic." *Chicago Schools Journal,* 12:26-29, September, 1929.
15. HUNKINS, R. V. and BREED, F. H. "The Validity of Arithmetical Reasoning Tests." *Elementary School Journal,* 23:453-66, 1923.

16. HYDLE, L. L. and CLAPP, FRANK L. *Elements of Difficulty in the Interpretation of Concrete Problems in Arithmetic.* Bureau of Educational Research Bulletin, No. 9. University of Wisconsin, 1927.
17. IMMACULATA, SISTER M. *A Critical Study of Arithmetic Reasoning Tests.* Educational Research Bulletin, Vol. 2, No. 6. The Catholic University of America, 1927.
18. KLINE, L. W. and ANDERSON, PHOEBE K. "The Rôle of Habit in Reasoning." *School Science and Mathematics,* 26:156-67, February, 1926.
19. KNIGHT, F. B. "Crucial Questions on Arithmetic Testing," *Chicago Schools Journal,* 11:4-9, 46-48, September and October, 1928.
20. LESSENGER, W. E. "Reading Difficulties in Arithmetical Computation." *Journal of Educational Research,* 11:287-91, April, 1925.
21. LEWIS, MARGUERITE, LANE, RUTH M., HOOVER, NELLE W., and KNIGHT, FANNY M. "Methods of Attacking Problems." *Educational Research Bulletin,* 13:283-86, October, 1924.
22. LIGDA, PAUL. "The Systematic Solution of Arithmetic Problems." *School Science and Mathematics,* 28:24-33, 172-80, January, February, 1928.
23. LUTES, O. S. *An Evaluation of Three Techniques for Improving Ability to Solve Arithmetic Problems.* University of Iowa Monographs in Education, No. 6, College of Education, University of Iowa, 1926.
24. MONROE, WALTER S. *How Pupils Solve Problems in Arithmetic.* University of Illinois Bulletin, Vol. 26, No. 23. Bureau of Educational Research Bulletin No. 44. University of Illinois, 1928.
25. MORTON, R. L. "Pupils' Errors in Solving Arithmetic Problems." *Educational Research Bulletin* (Ohio State University), 4:155-58, April, 1925.
26. MORTON, R. L. "An Evaluation of Arithmetic Teaching." *Educational Research Bulletin,* 8:267-73, September, 1929.
27. MORTON, R. L. "An Analysis of Errors in the Solution of Arithmetic Problems." *Educational Research Bulletin,* 4:187-90, April, 1925.
28. MORTON, R. L. "Solving Arithmetic Problems." *Educational Research Bulletin,* 4:203-07, May, 1925.
29. MYERS, C. C. *The Tragedy of Errors in Arithmetic.* Plymouth Press, Chicago.
30. MYERS, C. C. "Imagination in Arithmetic." *Journal of Education,* CV: 662-63, June, 1927.
31. NATIONAL COUNCIL OF TEACHERS OF MATHEMATICS. *Second Yearbook and Third Yearbook.* Bureau of Publications, Teachers College, Columbia University, 1928.
32. NATIONAL SOCIETY FOR THE STUDY OF EDUCATION. *Twenty-ninth Yearbook.* Public School Publishing Company, 1930.
33. NEWCOMB, R. S. "Teaching Pupils How to Solve Problems in Arithmetic." *Elementary School Journal,* Vol. 23, pp. 183-89, 1922-23.
34. OSBURN, W. J. *Reading Difficulties in Arithmetic,* State Department of Public Instruction, Madison, Wisconsin, 1925.

35. OSBURN, W. J. *Diagnostic and Remedial Treatment of Errors in Arithmetical Reasoning.* State Department of Public Instruction, Madison, Wisconsin, 1922 (mimeographed).

36. OTIS, ARTHUR S. "The Visual Method of Solving Arithmetic Problems," *Mathematics Teacher,* XXI: 483-89. December, 1928.

37. RIMER, LENORE E., WILSON, ANNA, and KNIGHT, H. B. "How Pupils Solve Problems in Arithmetic." *Educational Research Bulletin,* 13: 272-74, October, 1924.

38. ROLING, PEARL, BLUME, CLARA L., and MOREHART, MARG S. "Specific Causes of Failure in Arithmetic Problems." *Educational Research Bulletin,* 13:271-72, October, 1924.

39. SANFORD, VERA. "Extraneous Details." *Mathematics Teacher,* 21: 83-91, February, 1928.

40. SMITH, D. E. "The Question of Problems in Elementary Mathematics." *Educational Review,* 31:300-05.

41. STEVENSON, P. R. "Difficulties in Problem Solving." *Journal of Educational Research,* 11:95-103, February, 1925.

42. STEVENSON, P. R. "Increasing the Ability of Pupils to Solve Arithmetic Problems." *Educational Research Bulletin,* 13:267-71, October, 1924.

43. STONE, JOHN C. *Arithmetical Abilities and Some Factors Determining Them.* Contributions, No. 19. Bureau of Publications, Teachers College, Columbia University, 1908.

44. THORNDIKE, E. L. "Measurements of Ability to Solve Arithmetical Problems." *Pedagogical Seminary,* 21:495-503, 1914.

45. UPTON, C. B. "Changing the Curriculum in Arithmetic." *Teachers College Record,* 18:341-60, 1926.

46. WASHBURNE, CARLETON W. "Comparison of Two Methods of Teaching Pupils to Apply the Mechanics of Arithmetic to the Solutions of Problems." *Elementary School Journal,* 27:758-67, June, 1927.

47. WASHBURNE, CARLETON W. and MORPHETT, MABEL VOGEL. "Unfamiliar Situations as a Difficulty in Solving Arithmetic Problems." *Journal of Educational Research,* 18:220-24, October, 1928.

48. WASHBURNE, CARLETON, and OSBORNE, RAYMOND. Reports of Experiments by the Committee of Seven, in the *Elementary School Journal,* September, November, December, 1926, *et seq.*

49. WHEAT, HARRY E. *Relative Merits of Conventional and Imaginative Type of Problems in Arithmetic.* Contribution to Education, No. 359. Bureau of Publications, Teachers College, Columbia University, 1928.

50. WILSON, ESTALINE. "Improving the Ability to Read Arithmetic Problems." *Elementary School Journal,* 18:380-86, January, 1922.

51. WINCH, W. H. "Accuracy in School Children. Does Improvement in Numerical Accuracy 'Transfer'?" *Journal of Educational Psychology,* 1:557-89, December, 1910.

52. WINCH, W. H. "Further Work on Numerical Accuracy in School Children. Does Improvement in Numerical Accuracy Transfer?" *Journal of Educational Psychology,* 2:262-71, May, 1911.

APPENDIX

AN ANALYSIS OF INCORRECT SOLUTIONS TO THREE-STEP AND FOUR-STEP PROBLEMS

The results of this study on problem solving as given on pages 45 and 68 are based on the assumption made on page 2, namely, that on the *final test* if a pupil obtains the wrong answer in solving a given problem, it indicates inability to reason correctly in solving that problem, since the computational element had presumably been eliminated, to a reasonable extent, through the computational drills preceding the final test.

It seems desirable before closing this study to see to what extent the above assumption was justified. In other words, to what extent were computational errors still found in the final test, these errors causing problems to be marked wrong which were correct so far as reasoning is concerned.

The above question was answered by making an analysis of a random sampling of incorrect solutions appearing in the final test of three-step and four-step problems, particularly to determine the ratio of the number of incorrect solutions due to errors in computation to the total number of incorrect solutions made. The results of this analysis are found in Tables 10, 11, 12, and 13, on pages 76 and 77.

An analysis shows that of 843 incorrect solutions to the three-step problems made by beginning fourth-grade pupils in their final problem tests (see Table 10):

A.

(1) 175/843 or 20 per cent of the incorrect solutions examined were due to incorrect computation, with the arithmetical processes correctly selected.

(2) 371/843 or 44 per cent were due to the selection of wrong processes, but with correct computation.

(3) 205/843 or 25 per cent were due to both wrong computation and the selection of incorrect processes.

(4) 92/843 or 11 per cent were due to incomplete work.

In an analysis of 443 incorrect solutions to three-step problems made by beginning sixth-grade pupils in their final problem tests (see Table 11):

B.

(1) 118/443 or 26 per cent of the incorrect solutions examined were due to incorrect computation, with the arithmetical processes correctly selected.

TABLE 10

A SAMPLING ANALYSIS OF INCORRECT SOLUTIONS TO THREE-STEP PROBLEMS MADE BY BEGINNING *FOURTH GRADE PUPILS*

A	B	C	D	E
Problem	Correct process wrong computation	Wrong process correct computation	Wrong process wrong computation	Incomplete work
1.......	2	9	6	2
2.......	4	12	2	1
3.......	4	14	8	2
4.......	9	11	3	8
5.......	5	16	7	2
6.......	3	13	10	6
7.......	13	16	9	0
8.......	11	14	10	5
9.......	7	13	9	4
10.......	8	10	17	9
11.......	8	15	8	5
12.......	4	22	11	1
13.......	9	23	6	2
14.......	5	22	7	8
15.......	7	14	11	2
16.......	3	18	15	6
17.......	5	23	8	1
18.......	6	10	9	5
19.......	6	10	3	1
20.......	11	14	11	2
21.......	12	8	6	2
22.......	10	14	6	6
23.......	9	15	13	3
24.......	9	14	7	2
25.......	5	21	3	7
Total....	175	371	205	92

TABLE 11

A SAMPLING ANALYSIS OF INCORRECT SOLUTIONS TO THREE-STEP PROBLEMS MADE BY BEGINNING *SIXTH GRADE PUPILS*

A	B	C	D	E
Problem	Correct process wrong computation	Wrong process correct computation	Wrong process wrong computation	Incomplete work
1.......	1	11	0	1
2.......	2	0	2	2
3.......	7	1	6	2
4.......	0	1	1	10
5.......	10	1	0	3
6.......	11	3	10	7
7.......	4	2	2	8
8.......	6	1	1	3
9.......	7	3	3	6
10.......	2	7	3	7
11.......	1	4	1	13
12.......	3	4	1	6
13.......	4	5	2	7
14.......	12	2	6	8
15.......	6	5	2	8
16.......	5	8	5	14
17.......	3	1	3	8
18.......	2	6	3	7
19.......	1	3	1	7
20.......	5	2	9	6
21.......	3	0	1	6
22.......	9	1	3	11
23.......	2	2	0	4
24.......	5	1	4	7
25.......	7	5	7	15
Total....	118	74	75	176

Table 10 is to be interpreted as follows:— In problem one of the final test, two pupils missed this problem because of their error in computation. Their reasoning was correct, having selected the correct processes, but they had not computed accurately. Nine pupils missed problem one because they had not reasoned correctly. All of their computations were correct, but they had failed to select the right processes. Six pupils missed problem one because of both incorrect computation and the selection of wrong processes. Two pupils missed problem one because they were at a loss in knowing what to do. Their work was incomplete. Tables 11, 12, 13 are to be interpreted in the same manner.

TABLE 12

A Sampling Analysis of Incorrect Solutions to Four-Step Problems Made by Beginning *SIXTH GRADE PUPILS*

A	B	C	D	E
Problem	Correct process wrong computation	Wrong process correct computation	Wrong process wrong computation	Incomplete work
1.......	26	5	11	9
2.......	14	6	6	9
3.......	18	3	4	6
4.......	9	16	5	11
5.......	12	5	4	6
6.......	14	2	1	4
7.......	3	12	5	13
8.......	7	8	7	2
9.......	16	15	7	3
10.......	8	14	5	2
11	0	3	2	13
12.......	4	7	9	15
13.......	10	12	7	12
14.......	10	1	7	8
15.......	10	8	7	16
16.......	5	4	0	1
17.......	3	30	11	5
18.......	5	4	1	1
19.......	15	2	4	3
20.......	9	2	1	6
21.......	18	13	8	10
22.......	3	34	10	2
23.......	4	16	6	5
24.......	3	19	9	10
25.......	14	1	5	3
Total....	230	232	133	175

TABLE 13

A Sampling Analysis of Incorrect Solutions to Four-Step Problems Made by Beginning *SEVENTH GRADE PUPILS*

A	B	C	D	E
Problem	Correct process wrong computation	Wrong process correct computation	Wrong process wrong computation	Incomplete work
1.......	46	6	3	17
2.......	35	2	0	9
3.......	35	1	0	2
4.......	24	9	4	19
5.......	24	3	0	8
6.......	26	0	0	0
7.......	9	31	1	23
8.......	17	7	0	2
9.......	49	16	2	10
10.......	19	15	0	0
11.......	2	10	5	37
12.......	27	7	3	28
13.......	26	9	6	15
14.......	33	0	2	5
15.......	26	4	5	13
16.......	14	0	0	0
17.......	13	36	4	6
18.......	16	1	0	5
19.......	41	4	4	8
20.......	28	4	2	1
21.......	35	0	8	28
22.......	4	36	2	6
23.......	4	12	4	8
24.......	21	16	2	16
25.......	25	1	0	1
Total....	599	230	57	267

(2) 74/443 or 17 per cent were due to the selection of wrong processes, but with correct computation.

(3) 75/443 or 17 per cent were due to both wrong computation and the selection of incorrect processes.

(4) 176/443 or 40 per cent were due to incomplete work.

An analysis of 770 incorrect solutions to four-step problems made by beginning sixth-grade pupils in their final problem tests (see Table 12):

C.

(1) 230/770 or 30 per cent of the incorrect solutions examined were due to incorrect computation, with the arithmetical processes correctly selected.

(2) 232/770 or 30 per cent were due to the selection of wrong processes, but with correct computation.

(3) 133/770 or 17 per cent were due to both wrong computation and the selection of incorrect processes.

(4) 175/770 or 23 per cent were due to incomplete work.

The following analysis shows that of 1153 incorrect solutions to four-step problems made by beginning seventh-grade pupils, in their final problem tests (see Table 13):

D.

(1) 599/1153 or 52 per cent of the incorrect solutions examined were due to incorrect computation, with the arithmetical processes correctly selected.

(2) 230/1153 or 20 per cent were due to the selection of wrong processes, but with correct computation.

(3) 57/1153 or 5 per cent were due to both wrong computation and the selection of incorrect processes.

(4) 267/1153 or 23 per cent were due to incomplete work.

Referring to Table 3 on page 46, it will be seen that the entire Grade 4 was *successful* on 40 per cent of the three-step problems in the final test, which means that they *failed* on 60 per cent of the problems per pupil. According to the above data (Item A-1) 20 per cent of all the incorrect problems were marked incorrect because of incorrect computation though they were correct as to reasoning. Hence 20 per cent of the group of 60 per cent mentioned above were really *correct* as to reasoning. If we take 20 per cent of 60 per cent we get 12 per cent. Subtracting 12 per cent from 60 per cent we have 48 per cent. Hence only 48 per cent of the three-step problems in Grade 4 were solved incorrectly so far as reasoning is concerned. This means that 52 per cent of the three-step problems were really *correct as to reasoning.* Hence we can say that on the basis of correct reasoning fourth-grade pupils solved three-step problems with 52 per cent mastery in place of 40 per cent mastery, as indicated in Table 3.

The following tabulation shows the per cents taken from Table 3 and corrected as to correct reasoning by data in items A1, B1, C1, and D1, according to the procedure indicated above.

Grade	Per Cent of Problems Solved Correctly Per Pupil (from Table 3)		Corrected Per Cent on Basis of Correct Reasoning	
	Three-Step	Four-Step	Three-Step	Four-Step
Grade 4	40%	....	52%	....
Grade 6	56%	44%	67%	61%
Grade 7	....	56%	....	78%

Among the outcomes of this analysis the following seem to be especially important:

(1) Conclusions regarding a pupil's ability to reason in one-, two-, three-, and four-step problems, based alone on the answer obtained, are likely to be very misleading.

(2) From the teacher's standpoint satisfactory work in problem solving, which means both correct answers and correct reasoning, is very closely related to computational ability.

(3) The tendency in certain schools, especially in those where considerable attention is given to projects and activities, to decrease the amount of time given to computation, seems destined to lead to disappointing results.

(4) The drill upon computation in Grades 5, 6, and 7 must not be relaxed if satisfactory conditions in problem solving are to be achieved.

From an analysis of incorrect solutions similar to that made in Tables 10, 11, 12, and 13, teachers may obtain useful facts for directing the proper teaching emphasis in the matter of problem solving or for directing their classroom procedure. An examination of Table 10 shows that after the final test had been given, the major emphasis needed to be placed on training the pupils to select the right processes. Just what may be the nature of the specific causes for failure of pupils to select the right processes cannot be determined from this analysis. However, a further analysis of this particular weakness may reveal vocabulary difficulties, faulty teaching techniques, reading weaknesses, lack of proper symbol interpretations, etc., or a general lack of understanding on the part of the teacher or pupil, or both, as to what constitutes reasoning.

TABLE 14

PROBLEM PERFORMANCE

PER CENT OF ALL PUPILS SOLVING EACH PROBLEM CORRECTLY IN THE PRELIMINARY TEST

CITIES	TOTAL NO. WORKING ALL PROB.	PER CENT OF PUPILS SOLVING EACH PROBLEM CORRECTLY																								
		1	2	3	4	5	6	7	8	9	10	11	12	13	14	15	16	17	18	19	20	21	22	23	24	25
GRADE 3		*One-Step Problems*																								
Champaign, Ill.	261	61	74	37	46	54	30	66	41	9	46	61	35	45	48	7	33	26	34	26	25	41	33	69	32	35
Teaneck, N. J.	181	54	75	44	54	60	38	64	50	9	39	70	40	55	58	10	37	21	44	30	33	45	40	79	42	33
		Two-Step Problems																								
Champaign, Ill.	261	66	46	51	58	62	21	15	39	64	26	9	6	41	11	51	44	57	28	48	52	43	57	39	51	48
Teaneck, N. J.	181	60	42	43	51	54	18	20	41	57	20	6	5	42	7	44	41	51	27	38	40	37	51	35	48	41
GRADE 4		*One-Step Problems*																								
Champaign, Ill.	218	81	76	77	75	70	68	68	61	59	51	64	52	54	59	70	68	70	70	40	41	68	68	58	50	74
Teaneck, N. J.	196	85	76	78	80	70	71	74	68	64	60	70	60	60	62	73	68	74	74	49	47	75	71	67	55	79

Two-Step Problems

Champaign, Ill.	218	54	54	53	52	42	42	37	27	41	20	17	27	41	32	32	33	48	13	22	10	53	35	21	13	19
Teaneck, N. J.	196	56	55	60	56	46	49	45	29	47	29	20	29	43	35	43	37	54	21	26	13	69	57	31	11	19

Three-Step Problems

Champaign, Ill.	218	62	38	39	19	65	8	30	30	26	11	30	25	22	15	24	6	23	12	41	15	20	18	16	18	12
Teaneck, N. J.	196	67	49	41	20	67	7	34	38	30	15	37	32	20	14	28	4	17	16	54	15	23	21	18	20	23

Four-Step Problems

Champaign, Ill.	218	22	27	16	36	28	18	10	19	22	2	18	19	2	15	17	11	32	31	24	12	14	4	9	13	10
Teaneck, N. J.	196	30	32	14	40	33	19	11	17	23	2	15	16	4	15	18	17	40	41	24	10	21	2	4	11	18

GRADE 5

One-Step Problems

Champaign, Ill.	192	74	85	48	78	78	84	82	83	78	72	46	36	57	58	49	65	67	52	76	52	46	55	67	42	60
Teaneck, N. J.	111	80	80	52	80	81	86	86	84	85	80	54	37	65	67	53	70	79	61	84	55	41	75	76	50	74

Two-Step Problems

Champaign, Ill.	192	49	29	36	21	47	52	43	68	31	26	49	44	51	53	23	28	41	20	23	23	42	42	15	15	21
Teaneck, N. J.	111	62	31	41	28	57	59	49	70	28	32	60	60	59	60	24	37	59	24	44	25	50	44	18	24	26

TABLE 14—(*Continued*)

Cities	Total No. working all prob.	Per cent of pupils solving each problem correctly																								
		1	2	3	4	5	6	7	8	9	10	11	12	13	14	15	16	17	18	19	20	21	22	23	24	25
		Three-Step Problems																								
Champaign, Ill.	192	33	56	52	30	24	10	30	25	21	20	11	10	24	10	14	10	33	39	33	23	30	30	31	22	5
Teaneck, N. J.	111	37	60	62	40	29	10	30	31	25	29	11	13	27	13	10	16	37	40	32	20	43	40	40	42	5
		Four-Step Problems																								
Champaign, Ill.	192	16	6	15	15	25	10	10	18	11	25	33	15	3	16	6	46	8	31	23	32	9	10	18	2	18
Teaneck, N. J.	111	10	6	20	21	35	14	18	14	10	20	40	25	5	20	7	36	10	40	30	40	12	17	23	2	21
GRADE 6		*One-Step Problems*																								
Champaign, Ill.	255	85	93	82	91	91	89	91	86	76	83	64	54	71	73	57	78	76	71	83	58	61	74	82	55	68
Teaneck, N. J.	148	75	93	91	88	92	86	91	91	84	84	70	68	74	70	66	77	83	69	86	62	55	78	80	59	67
		Two-Step Problems																								
Champaign, Ill.	255	57	29	30	25	63	68	43	87	50	34	70	71	62	68	31	45	59	38	46	11	46	58	24	24	45
Teaneck, N. J.	148	60	34	36	33	58	70	36	80	52	39	64	70	64	70	40	43	70	40	36	17	39	49	32	34	44

Three-Step Problems																										
Champaign, Ill.	255	58	61	53	48	37	15	26	35	35	12	15	24	44	20	38	26	46	55	44	34	51	50	54	39	20
Teaneck, N. J.	148	55	64	55	41	41	16	17	39	29	10	17	22	38	14	28	22	46	37	37	22	48	40	41	29	14
Four-Step Problems																										
Champaign, Ill.	255	10	23	36	27	39	43	14	26	11	27	33	12	14	28	23	72	9	55	25	35	11	6	22	2	35
Teaneck, N. J.	148	8	26	28	23	36	49	18	33	14	32	30	19	10	24	15	65	5	51	28	41	8	5	23	3	35
GRADE 7																										
One-Step Problems																										
Champaign, Ill.	268	81	91	90	91	89	90	90	90	85	89	79	70	75	81	66	84	85	82	88	74	79	85	86	75	76
Teaneck, N. J.	168	84	94	91	93	93	94	91	94	70	90	71	72	65	85	69	91	85	81	85	87	75	76	90	85	81
Two-Step Problems																										
Champaign, Ill.	268	80	54	40	55	72	75	54	88	71	52	84	82	75	79	54	59	76	50	53	32	60	73	44	49	60
Teaneck, N. J.	168	78	51	47	60	76	80	46	88	80	63	89	85	68	81	60	65	81	57	60	24	65	72	40	59	72
Three-Step Problems																										
Champaign, Ill.	268	74	75	61	61	47	34	35	54	54	43	33	52	60	36	46	48	76	63	71	55	78	60	73	59	40
Teaneck, N. J.	168	74	78	70	70	47	41	40	60	53	31	30	49	71	47	60	56	80	68	78	66	82	71	80	59	41

TABLE 14—*Continued*

Cities	Total No. working all prob.	Per cent of pupils solving each problem correctly																								
		1	2	3	4	5	6	7	8	9	10	11	12	13	14	15	16	17	18	19	20	21	22	23	24	25
		Four-Step Problems																								
Champaign, Ill.	268	22	43	49	38	67	66	25	44	32	43	33	25	41	48	43	80	22	69	48	64	29	31	49	13	66
Teaneck, N. J.	168	16	50	57	40	71	82	26	43	21	48	34	34	37	44	40	81	16	71	40	69	24	28	50	9	63

TABLE 15

PROBLEM PERFORMANCE

Per cent of All Pupils Solving Each Problem Correctly in the Preliminary Test

GRADE 3																										
		One-Step Problems																								
Both cities	442	57	74	40	50	57	34	65	45	9	42	65	37	50	53	8	35	23	39	28	29	43	36	74	37	34
		Two-Step Problems																								
Both cities	442	63	44	47	54	58	19	17	40	60	23	7	5	41	9	47	42	54	27	43	46	40	54	37	49	44

GRADE 4		*One-Step Problems*																								
Both cities	414	83	76	77	77	70	69	71	64	61	55	67	56	57	60	71	68	72	72	44	44	71	69	62	52	76
		Two-Step Problems																								
Both cities	414	55	54	56	54	40	45	41	28	44	24	18	28	42	33	37	35	51	17	24	11	61	46	26	12	19
		Three-Step Problems																								
Both cities	414	64	43	40	19	66	7	32	34	28	13	33	28	21	14	26	5	20	14	47	15	21	19	17	19	17
		Four-Step Problems																								
Both cities	414	26	29	15	38	30	18	10	18	22	2	16	17	3	15	17	14	36	36	24	11	17	3	6	12	14
GRADE 5		*One-Step Problems*																								
Both cities	303	77	82	50	79	79	85	84	83	81	76	50	36	61	62	51	67	73	56	80	53	43	65	71	46	67
		Two-Step Problems																								
Both cities	303	55	30	38	24	52	55	46	69	29	29	54	52	55	56	23	32	50	22	33	24	46	43	16	19	23

TABLE 15—(*Continued*)

Cities	Total No. working all prob.	Per cent of pupils solving each problem correctly																								
		1	2	3	4	5	6	7	8	9	10	11	12	13	14	15	16	17	18	19	20	21	22	23	24	25
		Three-Step Problems																								
Both cities	303	35	58	57	35	26	10	30	28	23	24	11	11	25	11	12	13	35	39	32	21	36	35	35	32	5
		Four-Step Problems																								
Both cities	303	13	6	17	18	30	12	14	16	10	22	36	20	4	18	6	41	9	35	26	36	10	13	20	2	19
GRADE 6		*One-Step Problems*																								
Both cities	403	80	93	86	89	91	87	91	88	80	83	67	61	72	71	61	77	79	70	84	60	58	76	81	57	67
		Two-Step Problems																								
Both cities	403	58	31	33	29	60	69	39	83	51	36	67	70	63	69	35	44	64	39	41	14	42	53	28	29	44

Three-Step Problems

Both cities	403	56	62	54	44	39	15	21	37	32	11	16	23	41	17	33	24	46	46	40	28	49	45	47	34	17

Four-Step Problems

Both cities	403	9	24	32	25	37	46	16	29	12	29	31	15	12	26	19	68	7	53	26	38	9	5	22	2	35

GRADE 7

One-Step Problems

Both cities	436	82	92	90	92	91	92	90	92	77	89	75	71	70	83	67	87	85	81	86	80	77	80	88	80	78

Two-Step Problems

Both cities	436	79	52	43	57	74	77	50	88	75	57	86	83	71	80	57	62	78	53	56	28	62	72	42	54	66

Three-Step Problems

Both cities	436	74	76	65	65	47	37	37	57	53	37	31	50	65	41	53	52	78	65	74	60	80	65	76	59	40

Four-Step Problems

Both cities	436	19	46	53	39	69	74	25	43	26	45	33	29	39	46	41	80	19	70	44	66	26	29	49	11	64